The Deployment Life Study

Methodological Overview and Baseline Sample Description

Terri Tanielian, Benjamin R. Karney, Anita Chandra, Sarah O. Meadows, and the Deployment Life Study Team

Prepared for the United States Army and the Office of the Secretary of Defense

The research described in this report was prepared for the United States Army and the Office of the Secretary of Defense. The research was conducted jointly within the RAND Arroyo Center and the National Defense Research Institute under Contracts W74V8H-06-C-0001 and W74V8H-06-C-0002.

Library of Congress Cataloging-in-Publication Data is available for this publication.

ISBN: 978-0-8330-7992-3

The RAND Corporation is a nonprofit institution that helps improve policy and decisionmaking through research and analysis. RAND's publications do not necessarily reflect the opinions of its research clients and sponsors.

Support RAND—make a tax-deductible charitable contribution at www.rand.org/giving/contribute.html

RAND OFFICES

SANTA MONICA, CA • WASHINGTON, DC

PITTSBURGH, PA • NEW ORLEANS, LA • JACKSON, MS • BOSTON, MA

CAMBRIDGE, UK • BRUSSELS, BE

www.rand.org

The Deployment Life Study Team

Principal Investigators

Terri L. Tanielian
Benjamin R. Karney
Anita Chandra
Sarah O. Meadows

Project Director

Karen Yuhas

Co-Investigators

Jennifer L. Cerully
Beth Ann Griffin
Lisa H. Jaycox
Margaret C. Harrell
Stacy Ann Hawkins[a]
Sandraluz Lara-Cinisomo[a]
Laurie T. Martin
Sebastian Negrusa[a]
Terry L. Schell
Thomas E. Trail
Wendy M. Troxel

Research Assistance

Emily M. Gillen[a]
Racine S. Harris
Phoenix Voorhies

Survey Data Collection, Management, and Programming Support

Robin Beckman
Bernadette Benjamin
Tania Gutsche
Josephine Levy
Regan Main
Adrian Montero
Julie Newell
Janet M. Hanley
Teague Ruder
Alexandria Smith
Albert Weerman

Administrative Support

Emily Bever[a]
Stacy Fitzsimmons
Keeley Judd
Donna White

[a] No longer at RAND.

Preface

In the past decade, U.S. military families have experienced increased deployment tempo as U.S. soldiers, sailors, airmen, and marines have been deployed to hostile territory for extended and repeated durations. Therefore, policymakers and U.S. Department of Defense (DoD) leadership have placed an emphasis on family readiness for deployment and other military-related stressors. However, family readiness is not a well-understood construct.

In March 2009, the U.S. Army Surgeon General asked the RAND Corporation to design and conduct a longitudinal study of Army families to examine family readiness. The study would survey families at frequent intervals throughout a complete deployment cycle—that is, before a service member deploys (sometimes months before), during the actual deployment, and after the service member returns (possibly a year or more after he or she has redeployed). It would assess outcomes over time, including the following:

- the emotional, behavioral, and physical health of family members
- the quality of marital and parental relationships
- child well-being (e.g., school performance, social development)
- career outcomes (e.g., attitudes toward military service, retention intentions)
- family financial well-being.

In August 2009, the Defense Centers of Excellence for Psychological Health and Traumatic Brain Injury (DCoE) asked RAND to expand the study to include the other services: Navy, Air Force, and Marine Corps. This combined effort, called the Deployment Life Study, uses a single design and the same survey instruments (modified only slightly to make them service and component appropriate), thus allowing for potential comparisons across services and components (active and reserve).

Broadly, the Deployment Life Study is designed to address several policy questions. First, how are deployments associated with family well-being and overall functioning? Second, what family- and individual-level factors can account for both positive and negative adaptation to deployment (i.e., what constitutes family readiness)? And third, what policies and programs can DoD develop to help families navigate the stress associated with deployment?

This report details the theoretical model that informed the design of the Deployment Life Study, the content of the baseline assessment, the design and procedures associated with data collection, and sampling and recruiting procedures and provides a brief description of the baseline sample of military families. This report will be of interest to policy and program officials working on issues related to deployment cycle support and military family programming. The report may also be of interest to other researchers working on similar topics.

This research was jointly sponsored by the Office of the Surgeon General, U.S. Army, and by DCoE and conducted jointly within the RAND Arroyo Center's Army Health Program and the Forces and Resources Policy Center of the RAND National Defense Research Institute. RAND Arroyo Center, part of the RAND Corporation, is a federally funded research and development center sponsored by the United States Army. The RAND National Defense Research Institute is a federally funded research and development center sponsored by the Office of the Secretary of Defense, the Joint Staff, the Unified Combatant Commands, the Navy, the Marine Corps, the defense agencies, and the defense Intelligence Community.

The Project Unique Identification Code (PUIC) for the project that produced this document is DASGP09144.

For more information on the RAND Forces and Resources Policy Center, see http://www.rand.org/nsrd/ndri/centers/frp.html or contact the director (contact information is provided on the web page).

Contents

Figures and Tables

Figures

Tables

Summary

This report outlines the background, rationale, and methods of the Deployment Life Study; describes the baseline sample included in the study; and provides a context for future analyses using study data. The Deployment Life Study was motivated by a keen interest in military families on the part of U.S. policymakers and researchers. Specifically, recent attention by the U.S. Department of Defense (DoD) has focused on family readiness. When service members and their families are prepared for deployments, they should, in principle, be better equipped than other service members and families to maintain family cohesion and well-being across the deployment cycle. Yet exactly what it means for a military family to be "ready"—what ready families look like, what resources they use, what can help identify vulnerable families—is not well understood.

To address these questions, the Deployment Life Study is following 2,724 military families, including families from the Army, Air Force, Navy, and Marine Corps, both active and reserve, over a three-year period. The intent is to capture aspects of these families before deployment, during deployment, and after deployment. Interviews, either by Internet or by phone, occur every four months. Up to three household members are interviewed at each wave of data collection—the service member, the spouse of the service member, and a study child (if available). This survey design will collect an unprecedented amount of information about what military families look like and how they handle the challenges of military life, and deployments in particular.

Some generalities about the baseline sample of military families in the Deployment Life Study can be made. For example, the majority of service members in the Deployment Life Study sample are male and married to civilian wives. Most are in first marriages and have young children. Service members have an average of ten years of service. Thus, most (more than three-quarters) had experienced at least one prior deployment as of the baseline assessment. In general, according to studies of their civilian peers, levels of predeployment vulnerability in these families are low, as evidenced by the positively skewed distribution of known risk factors. At the time of the baseline survey, less than 10 percent of service members and spouses reported any incidence of several well-established vulnerability factors, e.g., marital conflict and violence, low-quality family environment, risky behaviors, drug and alcohol use, mental health problems. Nonetheless, the sample includes substantial variability across all of these variables, and that variability, along with changes that these families experience over time and over the deployment cycle, will provide the basis for future analyses.

Several strengths of the Deployment Life Study make it uniquely well suited to support future research on the antecedents, correlates, and consequences of family readiness among military families. First, the study design is prospective. Whereas most existing studies of military families rely on families to report on their experiences with deployment retrospectively, the Deployment Life Study assesses families before a deployment and then at multiple times during and after the deployment. The ability to compare postdeployment outcomes with a pre-deployment baseline is critical for disentangling the actual impact of a deployment from the impact of conditions or characteristics that may have existed before the deployment. That is, the prospective design supports precise estimates of the association between deployment and health and well-being outcomes among military families, controlling for preexisting conditions (that is, conditions existing before deployment).

Second, the Deployment Life Study is collecting data from multiple family members within the same families: not only service members but also their spouses and, in some cases, their children. Multi-informant studies offer a window into the role of agreement or disagreement within families, superior estimates of family cohesion, and the ability to correct for individual self-report biases when describing family outcomes.

Third, whereas many existing studies of military families focus on a narrow set of variables, the assessment strategy of the Deployment Life Study draws on a conceptual framework incorporating a wide range of domains, including demographic and personal history, military experiences, family and household composition, marital and parenting relationships, health and well-being, use of services, and adaptive and coping processes. Gathering data on all of these domains will allow analyses to describe the spread of deployment effects across domains, the sequence of deployment effects across domains, and the moderators and mediators of deployment effects on each family member.

Despite these many strengths, some limitations of the study should be noted. First, given difficulties with sampling, it could be that those families that could not be reached are, in some important way, unlike those families that did complete the baseline survey. Second, as with any survey that touches on sensitive topics (e.g., drug and alcohol use, mental health), we cannot be certain that respondents are always completely truthful. Third, our ability to make truly causal inferences about the effect of deployment on military families may be limited inasmuch as the survey cannot, and does not purport to, measure every possible cause for the observed outcomes. And fourth, like in every longitudinal study, sample attrition will occur that cannot be predicted by any characteristic (or set of characteristics) of families.

Many methodology and statistical techniques exist to address these limitations and will be used in future analyses that explore associations between the outcomes of interest and deployment.

Acknowledgments

Over the course of this study, we have benefited from the guidance and support of several individuals from our sponsoring offices. We wish to acknowledge our past and current project monitors and staff from the U.S. Army Office of the Surgeon General:[1] COL (Ret.) Elspeth Cameron Ritchie; COL Jonathan Jaffin; COL Paul Cordts; COL Rebecca Porter; LTC Christopher Ivany; and CDR Kathleen Watkins. Likewise, we thank and acknowledge the former and current project monitors and supporting staff from the Defense Centers of Excellence for Psychological Health and Traumatic Brain Injury (DCoE): CAPT Edward Simmer; Col Christopher Robinson; CAPT Paul Hammer; and CAPT Wanda Finch. The support of these individuals was instrumental in making our study happen. We also thank Colanda Cato, Evette Pinder, and Jeffrey Rhodes of DCoE's Deployment Health Clinical Center (DHCC) for their comments on an earlier draft of the document.

We also acknowledge the data-collection support from our colleagues at Abt SRBI, whose efforts led to a smooth data-collection process.

Through the peer-review process, we gained constructive and helpful comments and suggestions from Beth Asch and Gregory Leskin. We thank them for their time.

Finally, we want to thank the many military families who have participated, and who continue to participate, in the Deployment Life Study. We hope that, in some way, our work can contribute to their health, happiness, and well-being.

[1] Ranks and affiliations are current as of February 2013.

Abbreviations

ACE	Adverse Childhood Experiences
Army STARRS	Army Study to Assess Risk and Resilience in Servicemembers
BSF	Blue Star Families
BUPERS	Bureau of Naval Personnel
CATI	computer-assisted telephone interview
CONUS	continental United States
CTS-2	Conflict Tactic Scale 2
DCoE	Defense Centers of Excellence for Psychological Health and Traumatic Brain Injury
DMDC	Defense Manpower Data Center
DoD	U.S. Department of Defense
DoDI	Department of Defense instruction
FES	Family Environment Scale
FFCWS	Fragile Families and Child Wellbeing Study
FPNMAD	Florida Project on Newlywed Marriage and Adult Development
FRG	Family Readiness Group
FY	fiscal year
G-1	personnel management
IOM	Institute of Medicine
MSI	Marital Status Inventory
NHES	National Household Education Survey
OCONUS	outside the continental United States
OEF	Operation Enduring Freedom

OIF	Operation Iraqi Freedom
OND	Operation New Dawn
PANAS	Positive and Negative Affect Schedule
PCL	PTSD [posttraumatic stress disorder] Checklist
PCL-S	specific PTSD [posttraumatic stress disorder] Checklist
PCS	permanent change of station
PDRS	Post-Deployment Reintegration Scale
PHQ-8	eight-item Patient Health Questionnaire
PHQ-A	Patient Health Questionnaire for Adolescents
POC	point of contact
PSID	Panel Study of Income Dynamics
PTSD	posttraumatic stress disorder
SCARED	Screen for Child Anxiety Related Emotional Disorders
SDQ	Strengths and Difficulties Questionnaire
SF-12	12-item Short Form Health Survey
SOFS	Status of Forces Survey
SRG	RAND Survey Research Group
SSMP	Sample Survey of Military Personnel
TBI	traumatic brain injury

CHAPTER ONE

Introduction

> [W]hat I want and all my days I pine for is to go back to my house and see my day of homecoming. And if some god batters me far out on the wine-blue water, I will endure it, keeping a stubborn spirit inside me, for already I have suffered much and done much hard work on the waves and in the fighting. So let this adventure follow.
> —Homer, *The Odyssey of Homer*, Richmond Lattimore, trans., Book 5, lines 219–224

Of the many demands that military life imposes on service members and their families, deployments are perhaps the most stressful. When service members are required to leave their homes for extended periods, they and their spouses and children must adapt to new roles and responsibilities, even as they miss the daily support and connection that most people expect from family life. The demands do not end when the service member returns home. Deployments to theaters of operations pose additional challenges for military families. Experiences in battle can leave visible and invisible wounds that make reintegration to family life as much of an adjustment as being away (Tanielian and Jaycox, 2008). For thousands of years, military history and literature have recognized the challenges of departure, separation, and return. The enduring response to Homer's *Odyssey*, for example, speaks to the timelessness of a warrior's struggles to return to his wife and son after years away. In the modern era, sociologist Reuben Hill documented the difficulties some veterans faced reconnecting to their families after returning from World War II (Hill, 1949). Across time, the aspirations of warriors and their families have not changed much. In survey after survey, service members and their spouses continue to rate deployments as the single most stressful aspect of military life (Rosen and Durand, 2000).[1]

Since 2001, more than 2.5 million service members have been deployed to support Operation Iraqi Freedom (OIF), Operation Enduring Freedom (OEF), and Operation New Dawn (OND). These deployments have been longer and more frequent, with shorter rest periods between deployments, than deployments in previous conflicts over the past several decades (Belasco, 2007; Bruner, 2006; Serafino, 2003). Furthermore, deployed troops are surviving wounds at higher rates than ever before, resulting in more service members returning home with injuries and exposure to trauma (Sollinger, Fisher, and Metscher, 2008). As the consequences of these experiences began to emerge, the mainstream media started to pay attention.

[1] At the same time, for many military families, deployment has positive aspects (Hosek, Kavanagh, and Miller, 2006). Many service members look forward to deployment and see it as the realization of why they joined the military in the first place. Combat pay and other financial incentives can increase family financial well-being, especially among reservists (see Hosek, 2011). And some studies have found that deployment actually increases retention (see Hosek, 2011). Thus, despite the fact that deployment is the most frequently cited stressor among military families, it is important not to overlook the benefits it may also convey.

Current military leaders recognize that the military is a community of families and have been developing policies accordingly. For example, during the wars of the past decade, the military began to devote substantial efforts to support the families of service members (Rostker, 2006). Today, the majority of service members are married (Karney and Crown, 2007), and service members marry at higher rates than comparable civilians (Karney, Loughran, and Pollard, 2012). In this new context, the military provides health care and other services for more people out of uniform than in uniform (Hosek, Kavanagh, and Miller, 2006), including spouses and children. In an era of increased operational tempo, service members are subjected to the conflicting demands of their military and family commitments, which require loyalty, time, effort, and sacrifice (Segal, 1986). To alleviate this pressure and to support families, every service of the military has developed family resource centers (O'Keefe, Eyre, and Smith, 1984; Rubin and Harvie, 2012), and several programs address the challenges faced by families experiencing deployment in particular (e.g., the Yellow Ribbon Reintegration Program). Budgets for programs to support the health and well-being of service members and their families have never been higher. Nevertheless, in an environment of strained resources, assessment of which programs are most effective has become a key part of the discourse.

The explicit goal of these support efforts is to promote *family readiness*. The military invokes family readiness to recognize that the success of military operations abroad depends not only on the preparation of service members but also on the preparation of their families (Burnam et al., 1992; Schumm, Bell, and Resnick, 2001; Rosen and Durand, 2000). Families that have been prepared successfully (i.e., those that have high family readiness) presumably weather the stresses of deployment more successfully than families that are less prepared (i.e., those that have low family readiness). Thus, assessments of family readiness, prior to deployments, can play a role in allocating resources for families most likely to need them.

Yet, despite the centrality of family readiness as a rationale for deployment-related services, to date, the definition of the concept has been described in only the most-abstract terms. For example, the Marine Corps, per Marine Corps Order (MCO) 1754.9A (Commandant of the Marine Corps, 2012, p. 3), defines ready families as those that are "equipped with the knowledge and skills necessary to successfully meet the challenges of the military life." The specific skills and tools that serve these functions, however, have not been specified. Perhaps as a consequence of this general vagueness, spouses and service members report different definitions of *readiness*, ranging from emotional readiness to having their legal and financial documents in order (Werber et al., 2008).

In 2012, the U.S. Department of Defense (DoD) released DoD Instruction (DoDI) 1342.22, which defines family readiness as "the state of being prepared to effectively navigate the challenges of daily living experienced in the unique context of military service" (Under Secretary of Defense for Personnel and Readiness, 2012). Further, the DoDI states that ready families are knowledgeable about the potential challenges they may face, equipped with the skills to competently function in the face of such challenges, aware of the supportive resources available to them, and make use of the skills and supports in managing such challenges. Readiness, then, encompasses mobility and financial readiness, mobilization and deployment readiness, and personal and family life readiness. However, DoDI 1342.22 still does not precisely describe what specific knowledge and skills lead to family readiness, or what supportive resources lead to family readiness.

The need for a more precise understanding of the sources of readiness in the face of deployment has begun to be recognized. In the past several years, at least five congressional hearings (convened by the Senate Armed Services Committee, House Armed Services Committee, or the House Committee on Veterans' Affairs) and more than 60 bills have addressed the effects of deployment on military families. In 2009, the Institute of Medicine (IOM) was asked by Congress (§1661 of the National Defense Authorization Act [NDAA] for Fiscal Year [FY] 2008, Public Law 110-181, 2008) to undertake a review of the empirical evidence on the physical, mental health, and other readjustment needs of military families containing a member who had been deployed to OIF or OEF. Among the major findings of the report was documentation of the relative paucity of data adequate to support evidence-based policy on most issues of concern to OIF and OEF populations. Because the unique effects of deployment on family functioning have yet to be documented rigorously, the characteristics of families that are more or less successful at maintaining their functions across a deployment cycle are unknown as well. The report committee made several recommendations, including funding research on military families and deployments that addresses methodological and substantive gaps in the existing literature and that assesses effects of deployment across multiple domains (e.g., social and economic effects). Heeding this call, Congress has requested and required longitudinal studies of the effects of deployment on service members and veterans (see U.S. Senate, 2009; U.S. House of Representatives, 2009) and separate longitudinal studies of the effects of deployment on children of military personnel and families (see IOM, 2010, as one example).

The Deployment Life Study is designed to respond to these calls. The broad goal of the study is to gather data to support a refined understanding of family readiness. Which families are best able to withstand the strains of deployment, and what kinds of coping strategies characterize these families? Which families are most vulnerable to the negative consequences of deployment, and how might those families be targeted for extra support? The Deployment Life Study examines these questions by interviewing married service members, their spouses, and (when available) one child, at nine assessments spread across a three-year period, and across all phases of a deployment cycle. Each interview includes instruments measuring a wide range of variables relevant to understanding how the social, economic, and psychological well-being of military families changes across the deployment cycle. As a consequence of this breadth, the Deployment Life Study data set will ultimately be a resource for researchers from multiple disciplines and, most importantly, will provide robust data with which to assess the association between deployment and military family health and well-being. The specific aim of the current report is to document the rationale and methods of the Deployment Life Study, providing detail about the study design and measures and providing context for subsequent analyses.

Outline of This Report

The rest of this report is organized as follows. Chapter Two briefly reviews what we know about the association between deployment and the health and well-being of military families. The limitations of this research set the stage for the Deployment Life Study. Chapter Three briefly presents the theoretical model that informed the design of the Deployment Life Study and guided the content of the baseline assessment. Chapter Four presents the study approach, including details of the Deployment Life Study design and procedures, as well as sampling and recruitment. Chapter Five describes the constructs and instruments administered in the baseline assessment. Chapter Six describes the baseline sample. Chapter Seven provides a brief summary, discusses strengths and weakness of the Deployment Life Study, and offers a window into next steps. Finally, several appendixes, available online at www.rand.org/pubs/research_reports/RR209, provide copies of study documents, including the complete baseline assessment interview for service members, spouses, and children.

CHAPTER TWO

What We Know About Deployment and Military Families

Throughout U.S. history, the effects of deploying to war zones have been documented in the research literature, as well as in popular media. The effects of combat exposure on prior-era veterans brought increased understanding of mental health problems, such as posttraumatic stress disorder (PTSD). In addition, a large body of research has shown that having a family member deployed can have negative impacts on military spouses and children, although the bulk of this research has occurred in the past 20 years (for examples, see Flake et al., 2009; Jensen, Lewis, and Xenakis, 1986; Murphey, Darling-Churchill, and Chrisler, 2011; Norwood, Fullerton, and Hagen, 1996; Pincus et al., 2001; Segal, 1986; and Van Vranken et al., 1984). Although the stress of military deployments on families has long been recognized, prior research documenting the effects of deployment on the well-being of service members and their families has often been limited to postdeployment assessments, one service at a time, and a narrow set of health-related outcomes. These prior studies are important for setting the foundation for family support programs, but they have not addressed the effects that the pace and tempo of the current deployments can have on a range of family-related outcomes. As the pace and tempo of deployment over the past decade have increased, however, so did interest in the consequences of those sequential deployments for military families. This chapter briefly reviews the emerging research on deployments to Iraq and Afghanistan and their association with the health and well-being of service members, their spouses, and their children.

Deployment and Military Family Members

Deployment and Service Members

Deployments can threaten service members in two ways. Being separated from home can strain relationships with spouses and children, and exposure to combat while deployed can lead to physical injuries and trauma. To date, research on the effects of deployment for service members has focused primarily on the latter threat.

With respect to their physical health, service members who have been deployed experience higher rates of illness and worse physical functioning than those who have not been deployed, for several reasons. First, high numbers of military personnel have suffered physical injuries, including traumatic brain injury (TBI) from exposure to improvised explosive devices (IEDs) during deployments (Reger and Gahm, 2008). Second, some evidence suggests that, regardless of their direct exposure to combat, service members who have experienced a deployment face more pain and worse physical functioning than service members who have not experienced a deployment (Kline et al., 2010). Recent data from the U.S. Department of Veterans

Affairs have also shown that approximately 57 percent of the OEF and OIF veterans who visit a Veterans Health Administration facility have a diagnosis of a musculoskeletal problem (U.S. Department of Veterans Affairs, 2012). Third, the experience of deployment may also exacerbate risky behavior (e.g., participation in risky recreational activities, unprotected sex, illegal drug use), especially among deployed personnel with a predeployment history of engaging in such behavior (Thomsen et al., 2011). Fourth, deployed service members report an increased use of alcohol and tobacco, increasing their long-term disease risk (I. Jacobson et al., 2008; Smith et al., 2008).

Deployments have also been associated with negative consequences for service members' mental health outcomes. Some of these negative consequences are a direct result of physical injuries received during deployment. For example, experiencing combat-related injuries, such as TBI, can result in mood disorders and deficits in cognitive functioning (Vincent et al., 2008). Even in the absence of physical injury, however, combat exposure can leave service members vulnerable to developing PTSD and depression (Hoge, Castro, et al., 2004; Smith et al., 2008; Polusny et al., 2011; Ramchand et al., 2010; Schell and Marshall, 2008) and has been linked to higher rates of alcohol abuse as well (Thomas et al., 2010). Veterans who experience postdeployment mental health problems may also be at higher risk for other health-related problems. For example, Boscarino (2006) found that combat veterans with PTSD had elevated risks of cardiovascular mortality, external-cause mortality, and cancer mortality relative to combat veterans without PTSD. Of greatest current concern, the increase in mental health problems among deployed service members has been accompanied by a substantial increase in suicide and rates of suicidal ideation in service members. In the 2012 Blue Star Families (BSF) survey, for example, 9 percent of respondents reported knowing that the active-duty service member in the family had contemplated suicide at some point (BSF, 2012). Suicide rates have increased over time for service members, particularly within the Army in all settings, but are highest among those currently deployed (Army Study to Assess Risk and Resilience in Servicemembers [Army STARRS], 2012).

Generally, military deployment has a negative association with service member well-being and other outcomes. However, there are two areas in which deployment has actually been shown to have a positive impact, under certain circumstances, on earnings and military retention and promotion. First, deployments, with accompanying special and incentive pay (e.g., hazardous-duty pay), have been linked to an increase in earnings among the reserve component (Loughran, Klerman, and Martin, 2006; Martorell, Klerman, and Loughran, 2008). One possible explanation for an increase in earnings is that earnings collected while serving in a combat zone, which are typically higher than nondeployed earnings, are not taxed. Other research has found that enlistment in the military has a positive impact on later earnings, both while service members are still in the military and after they enter the civilian labor market (Loughran et al., 2011). Note that it is not clear how this earning effect is associated with deployments, although deployments are clearly a part of military service. Second, some evidence suggests that deployments are positively associated with retention, although these effects vary by service, component, and type of deployment (e.g., length, hostile versus nonhostile) (Fricker, Hosek, and Totten, 2003; Hosek and Martorell, 2009; Hosek and Miller, 2011).

Deployment and Spouses

Although the spouses of service members are not deployed themselves, their lives may be affected by a service member's deployment in at least three ways. First, their relationship to the service member may be strained by separation and anxiety and the resulting lack of contact for extended periods. Second, the nondeployed spouse must, while the service member is absent, take on additional responsibilities for household maintenance and often child care. Third, the nondeployed spouse, having adjusted to the service member's absence, must readjust upon the service member's return and frequently must cope with a partner who has suffered traumatic experiences while deployed. To date, research has emphasized the latter two of these effects.

With respect to the additional strains on the nondeployed spouse during deployment, multiple surveys confirm that spouses also experience deployment-related stress and that the strain of deployment is associated with negative mental health outcomes for other members of the family as well. For example, spouses whose partners are deployed report higher stress levels than the spouses of nondeployed service members (de Burgh, 2008), with 17 percent reporting seeking counseling for support during that time (BSF, 2012). A review of studies of female military spouses found that spouses of deployed service members reported higher rates of depression, anxiety disorders, sleep disorders, acute stress reaction, and adjustment disorders than those of wives of nondeployed service members (de Burgh et al., 2011). When a service member's deployment is unexpectedly extended, spouses report increased feelings of loneliness, anxiety, and depression, as well as declines in marital satisfaction (SteelFisher, Zaslavsky, and Blendon, 2008; Verdeli et al., 2011). A recent study of Army spouses noted an increase in mental health diagnoses and use of mental health services (Mansfield et al., 2010).

It might be expected that the stress of deployment would be relieved when the service member returns, but the postdeployment reintegration period is associated with its own challenges for spouses (Lara-Cinisomo et al., 2012). Three percent of the respondents in the BSF survey reported that the service member in the family had been diagnosed with TBI, and an additional 11 percent reported a PTSD diagnosis (BSF, 2012). One-quarter of spouses felt that their service members had displayed PTSD symptoms regardless of diagnosis, and 9 percent reported that the service member in the family had contemplated suicide (BSF, 2012). In instances in which service members reported more combat exposure, both partners scored higher on stress (Allen and Meyer, 1990). Renshaw (2011) studied the effect of service members' PTSD on spouses and relationships and found that the more that service members displayed symptoms of PTSD, the more distress their spouses reported and the lower their marital satisfaction.

Deployment and Children

The dramatic changes in family functioning that accompany deployment are likely to affect the well-being of some children in military families as well. Indeed, children of deployed parents are more likely to demonstrate internalizing and externalizing symptoms, such as anxiety, depression, and aggression (Aranda et al., 2011; Chartrand et al., 2008; Lester et al., 2010; Lincoln and Sweeten, 2011; Chandra, Lara-Cinisomo, Jaycox, Tanielian, Burns, et al., 2010; Chandra, Lara-Cinisomo, Jaycox, Tanielian, Han, et al., 2011). Children of deployed military parents are also at greater risk of experiencing psychosocial difficulties, including attention problems and problems in school (Aranda et al., 2011). One study linking children's medical records to their parent's deployment records observed an 11-percent increase in children's visits to doctors, a 19-percent increase in behavioral disorders, and an 18-percent increase in stress disorders when a military parent deployed (Gorman, Eide, and Hisle-Gorman, 2010). Perhaps as a consequence of the sudden lack of an additional parent to provide supervision, children with a deployed parent are also more likely to experience neglect or maltreatment (Gibbs et al., 2008).

With respect to academic outcomes, Engel, Hyams, and Scott (2006) found significant decreases in test scores for children of deployed parents, with longer deployments associated with lower test scores. Lyle (2006) found that the association of deployment with lower test scores was largest in younger children, those from single-parent homes, those with mothers in the Army, and those whose parents scored lower on the Armed Forces Qualification Test. Chandra, Martin, et al. (2010) conducted focus groups with school personnel who reported that, for many children, parental deployment led to sadness and anger, resulting in classroom disruption and negative effects on peer relationships. Gender differences reported by staff included more anger and aggressive behaviors for boys and more somatic complaints and internalizing in girls (Chandra, Martin, et al., 2010).

The Limits of Existing Research on Deployment Effects

Across the accumulated research on the consequences of deployments for military families, results suggest that those consequences are negative on many, but not all, outcomes that can be measured.[1] Before we accept this conclusion, however, several limitations of the existing research should be taken into account.

First, the existing research on deployment effects relies almost exclusively on retrospective designs. That is, researchers collected samples of military families and compared current outcomes in families that did or did not experience a prior deployment. This is a reasonable strategy for comparing the two groups, but it cannot distinguish between the consequences of deployment and problems that may have existed prior to deployment. Retrospective designs also offer no ability to estimate the degree of change that particular families experience across the deployment cycle, or the possibility of recovery over time after deployments. Further, retrospective designs may suffer from recall bias, which could systematically bias estimates (see Loughran, 2002). As a consequence, retrospective designs run the risk of exaggerating deployment effects and their duration.

Second, although the main effects of deployment have been mostly negative, this does not mean that most military families experience negative effects of deployment. Even if a particular negative outcome is more prevalent among families experiencing deployments than among other families, most military families may nevertheless avoid these negative consequences, and some may even thrive. Understanding the sources of variability in responses to deployment is crucial if military leaders are to tailor programs and allocate resources toward families that might need the resources most. Doing so requires research that emphasizes moderators of deployment effects (i.e., the qualities of families prior to deployment that account for how their well-being changes across the deployment cycle).

Third, although research to date has examined service members, spouses, and children, no large-scale studies have examined multiple respondents in the same families. As a result, the accumulated research offers no way of determining whether deployments have independent

[1] Again, deployment can have a positive association with some outcomes, including pay and retention (see Hosek, 2011), as well as personal growth and career advancement.

effects on each family member, or whether effects on one family member spread across the family system.

Fourth, although deployments are likely to affect a wide range of family outcomes, existing research has generally focused on a single domain at a time. The results of such research offer no way of knowing which outcomes may be affected first, which are directly versus indirectly affected by experiences during deployment, and which are affected more or less than any others. Understanding how military families react to deployment requires research that assesses the health, well-being, and characteristics of all members of military families, as well as relationships between spouses and their children, circumstances outside the family, and the varying demands of military service and other life stressors.

The Deployment Life Study

In light of the pressing need for data to inform policies to improve readiness in families facing deployments, and in light of the limitations of existing research on the effects of deployment, new research must describe the association between deployment and multiple dimensions of family well-being and examine how characteristics of families prior to deployment account for variability in their responses to the experience. In pursuit of this goal, the RAND Corporation is conducting the Deployment Life Study, a multiyear, longitudinal study to identify the antecedents, correlates, and consequences of family readiness across the deployment cycle. At baseline, the study recruited 2,724 Army, Air Force, Navy, and Marine Corps families. The sampling frame was restricted to married service members who were eligible for deployment within the next six to 12 months, and thus can be considered a sample of the deployable married military population. In each family, a service member, his or her spouse, and a child between the ages of 11 and 17 (if there was one in the household) each provided information independently. Interviewers obtained consent from each participating family member and conducted the baseline interviews by phone. Every four months thereafter, respondents log into the study website and complete a follow-up survey online. Data collection will include a total of nine waves spanning the entire deployment cycle, including periods of predeployment, deployment, and postdeployment.

The data accumulated through these efforts will be suitable for addressing a wide range of questions about the health and well-being of military families, but the following questions are priorities that guided the design of the study:

- *Controlling for initial conditions and functioning, how is deployment related to military family outcomes?* This study will evaluate the association between deployment and a range of outcomes, including those relevant to the functioning of the entire family unit and outcomes relevant to understanding the well-being of each individual family member. In particular, the multiwave longitudinal design will allow analyses to identify when in the deployment cycle specific needs are most pressing for families and the specific needs and consequences that arise most frequently at different stages of the deployment cycle.
- *Across families, what accounts for variability in how families react to deployment and the extent to which family outcomes return to baseline (i.e., predeployment) levels or better?* Of the factors accounting for such variability, which are the most important? The implications of deployment are likely to vary significantly across families, and even across individu-

als within families. The Deployment Life Study will assess and analyze the individual and family traits, resources, and circumstances that account for which families endure deployment well (or even benefit) and which families suffer.

- *What behaviors and programs best buffer families from any negative effects associated with deployment?* The project will examine the marital and family resources and coping behaviors that serve as mediators and moderators of the effects of deployment. Specific resources and processes will be assessed. The study will also be able to examine how changes in some outcomes (e.g., the marital relationship) are associated with changes in other outcomes (e.g., child school performance).

Analyses will focus on identifying the characteristics of more (or less) successful families over time, where *success* is defined in terms of a range of important outcomes that will be measured at every assessment. These include the following:

- the emotional, behavioral, and physical health of family members
- the quality of marital and parental relationships
- child outcomes (e.g., school performance, social development)
- military career outcomes (e.g., attitudes toward military service, retention intentions)
- financial well-being.

The Deployment Life Study is not the only project that has responded to the call for longitudinal research on military families. To facilitate comparisons between the Deployment Life Study and other prominent longitudinal studies that have also been examining this population, details about the designs of each study are presented side by side in Table 2.1. As the table reveals, the Deployment Life Study has the smallest sample of the five major studies of military families currently ongoing. Yet, as the table also reveals, the Deployment Life Study has several unique characteristics that will allow it to address issues that no other study can address as effectively.

Anchoring on Deployment

The primary rationale for conducting longitudinal research on military families is that longitudinal designs allow for estimates of changes in family outcomes after a deployment, controlling for preexisting differences between families prior to deployment (i.e., a first-difference approach).[2] The greatest benefits of this design, therefore, arise from the power to observe families before and after a deployment cycle. The Deployment Life Study is designed to take advantage of the power of such an analysis. Whereas the other ongoing longitudinal studies sampled from the general military population (some of whom may already be deployed and others of whom may never deploy over the course a study), the Deployment Life Study samples exclusively from the population of married service members eligible for deployment within the next six to 12 months. Moreover, to be eligible for the study, the service member could not be

[2] A difference-in-differences approach is often preferred to a first-difference approach, especially when one is interested in determining causal effects. Such an approach would compare pre- and postdeployment outcomes between families that did and did not experience deployment. At this time, it is unclear how many of the families in the study will not deploy during the observation period; so, a priori, it is not possible to determine whether the difference-in-differences approach is viable. Nonetheless, the analyses will utilize every means possible to rule out alternative causal scenarios and attempt to isolate the impact of deployment of military families.

Table 2.1
Comparing the Deployment Life Study and Other Studies of Military Families

Study Name	Sample Size	Population and Sampling Method	Service	Component	Number of Waves	Interval Between Waves	Constructs Assessed
Deployment Life Study	2,724	Stratified random sample of deployable married service members, spouses, and one child	Army, Navy,[a] Air Force, Marine Corps	Active, reserve	9 over 3 years	4 months	Emotional, behavioral, and physical health of family members; marital and parental relationships; military career outcomes; child outcomes; financial well-being
Military Family Life Project	28,552	Random sample of spouses of service members	Army, Navy, Air Force, Marine Corps	Active	4 over 2 years	6 months	Health and well-being; financial well-being; military life; deployment history; child outcomes; reunion and reunification
Millennium Cohort	100,000	Random sample of service members, and a spouse cohort	Army, Navy, Air Force, Marine Corps	Active, reserve	7 over 21 years	3 years	Physical and functional status; psychosocial assessment; medical conditions; self-reported symptoms; PTSD; alcohol and tobacco use; alternative medicine use; life events; occupational exposure
Army STARRS	1.6 million, archival data	Convenience sample of service members	Army	Active	5-year data-collection period	Not applicable; there are several data-collection efforts within Army STARRS, so no waves	Stress; deployment experiences; exposure to trauma; family and personal history; coping; personality and temperament; social networks and support
BSF 2012 Military Family Lifestyle Survey	4,234	Convenience sample of service members, their spouses, and their parents	Army, Navy, Air Force, Marine Corps	Active, reserve, veterans	3 to date, but not longitudinal with same sample	1 year	Pay, benefits, and changes to retirement; family well-being; satisfaction with military; deployment and wellness; financial security; military/civilian connectedness

[a] Because of the timeline required to secure the Navy sample, this enrollment started one year later than those for the other services. As a result, the Navy subsample will complete only seven total waves of data collection.

deployed at baseline. Thus, baseline assessment of the Deployment Life Study sample is a true baseline for a sample highly likely to experience a deployment over the course of the study.

Quarterly Assessments over Three Years

Military families may cope in different ways across different stages of the deployment cycle (i.e., functioning at predeployment may not be the same as functioning during deployment or postdeployment). Moreover, the consequences of deployments may themselves evolve or fade as a family adjusts and restores its predeployment equilibrium. Even longitudinal studies risk missing or mischaracterizing these effects if the interval between assessments is too long or the duration of the entire study is too short. By assessing families every four months for what is currently funded as a three-year study (nine assessments total), the Deployment Life Study is designed to produce higher-resolution descriptions of the deployment cycle than has been available in other studies. Because all families have been recruited prior to a deployment, not only will this design allow for assessments of family functioning at each stage of the deployment cycle; it will, in most cases, allow for *multiple* assessments at each stage, including multiple assessments shortly after a deployment. Thus, the Deployment Life Study will be able to describe how families change within each stage of the cycle, how long changes associated with deployment last, and what characteristics of families prior to deployment predict the length of a family's postdeployment return to equilibrium.

Data from Multiple Family Members

The vast majority of existing research on military families, and three of the five ongoing longitudinal studies described in Table 2.1, assess family functioning exclusively through the reports of a single family member, usually the service member or the spouse. There are several reasons that reliance on the reports of a single family member offer only a limited window into family functioning across a deployment cycle. First, service members and spouses may use different resources to cope with deployment. Second, coping strategies that are effective for one member of a family may not be effective for another. Third, to the extent that mental health is an outcome of major importance and the rates of mental health problems in this population are relatively high, the perceptions of any single family member may be distorted in ways that are impossible to estimate. Fourth, to the extent that one family member is asked to describe outcomes for another family member (e.g., parents reporting on their children, service members reporting on their spouses), associations among responses can be inflated by shared method variance.[3] The Deployment Life Study avoids these problems by obtaining responses from service members, their spouses, and one child between the ages of 11 and 17, if available. In addition, having multiple assessments of each family member will allow for the modeling of the spread of effects across family members over time.

The next chapter presents the conceptual model that framed the Deployment Life Study and helped guide the development of the survey instruments.

---

[3] That is, variance that is attributed to the method rather than to the constructs being measured. *Shared* refers to the fact that it applies to more than one person in the family.

CHAPTER THREE

Conceptual Model

In the interest of supporting a comprehensive understanding of the association between deployment and the health and well-being of military families, the Deployment Life Study includes assessments of a wide range of variables. Selection of those variables required a theoretical framework to suggest which ones, out of the countless constructs that could have been assessed, warranted inclusion within the limited claims we could make on the time of participating families. Thus, a preliminary step in the design of the Deployment Life Study was to identify a conceptual model that accounts for well-being in military families across the deployment cycle and so could offer direction for selecting constructs and forming initial hypotheses.

There were not many models from which to choose. Despite the substantial resources that have been directed toward supporting military families since the birth of the all-volunteer force in 1973, these programs have been characterized as "largely reactive, developed primarily in response to specific problems and their symptoms" (Bowen and Orthner, 1989, p. 180), rather than grounded in a priori theories of how military families function. Bowen and Orthner, in their edited volume, *The Organization Family: Work and Family Linkages in the U.S. Military*, describe the standards that a comprehensive theory must meet:

> There is a critical need for an explicit model of work-family linkages in the military (replete with underlying assumptions and operational outcome statements) that not only identifies the factors that promote level of adaptation to the multiplicity of organizational and family demands faced by service members and their families, but also specifies the direct and indirect impact that military policies, practices, and programs have on the ability of service members and their families to successfully respond to these demands. This model must reflect the dynamic and interactive quality of work and family life across the work and family life cycles. In addition, it must respect the tremendous age, ethnic, and cultural diversity found among families in the military services today by accounting for personal system-level influences, including the values, needs, and expectations of service members and their families toward both work and family life. Finally, for purposes of clinical and community intervention, the model must be practice based—capable of guiding the development, implementation, and evaluation of policies, programs, and practices in support of families. (p. 180)

Two decades after this call to arms, research on military families has not delivered a theory that meets these standards. In 2007, however, Karney and Crown published an integrative framework to account for success and failure in military marriages. Although goals of the Deployment Life Study extend beyond marital outcomes to include child outcomes, financial and employment outcomes, and the physical and mental health of each family member, the

Karney and Crown model provides a starting point for identifying general categories of constructs that may account for the outcomes of interest in the current study.

With the goal of understanding how families experience deployments, we can divide those general categories roughly into four supraordinate themes: preexisting conditions, deployment experiences, short-term outcomes, and long-term outcomes. These themes, and the general categories of constructs that fall within each of them, are described in more detail in this chapter. Many of the constructs included in the Deployment Life Study could be included under more than one theme, and in more than one category, depending on the research question and analysis. The examples we provide in this chapter are one categorization, but others are possible.

Preexisting Conditions

No family goes into a deployment as a blank slate. Rather, even before entering the predeployment phase, each military family possesses a unique set of traits, personal histories, capabilities, vulnerabilities, and resources that are likely to play a role in how family members function across the deployment experience and their readiness. Among the major categories of preexisting conditions are the following:

- *Enduring traits.* These are the relatively stable qualities of each family member, including demographic characteristics, personality, prior psychopathology, and personal history. Some qualities of family members, such as having more education, a stable childhood environment, and lack of personality problems or psychopathology (e.g., depression, anxiety), should contribute to readiness and effective adaptation across a deployment. Other qualities, such as a history of depression or substance abuse, should be associated with less-effective coping and poorer postdeployment outcomes.
- *Relationship resources.* These are attributes of the relationships among family members, including the duration of the marriage, the presence or absence of children, the number of children, and the quality of the relationships among all family members. Families with more relationship resources, such as a longer and higher-quality marriage, should endure the experience of deployment more successfully than families with fewer resources.
- *Nonmilitary circumstances.* Each family lives in its own ecological niche, characterized by specific neighborhoods, social networks, levels of chronic stress, and the availability of social support. Some niches, presumably those with more-abundant sources of support and lower levels of chronic stress, should promote more-resilient military families, and other niches, presumably those with fewer sources of support and more sources of stress, should predict families being more vulnerable to experiencing problems when confronted with the additional stresses of a deployment.
- *Prior military experiences.* Because this is a model to account for a family's experiences across a deployment, it makes sense to highlight preexisting conditions relating specifically to a family's experiences in the military prior to a deployment. Families differ by the rank and branch of the service member. Some families face deployment for the first time, whereas others have already experienced multiple deployments. Some families have long histories in the military and may come from military families, whereas others are new to the military and to the expectations of the institution. Families with deeper ties to the

military and longer histories in the military may cope more effectively with the strains of deployment than families for whom those strains are new or unexpected.

Experiences During Deployment

It is a mistake to use the word *deployment* and assume that the experience is the same for every service member who is deployed. On the contrary, the aftermath of deployments for military families is likely to depend on what happens during the deployment, and those experiences are likely to vary widely across families. Some categories of deployment experiences we identified are as follows:

- *Deployment characteristics.* Across service members, deployments may differ in length, time since the last deployment, and likelihood of exposure to combat. Some service members are deployed with their own units, and some are deployed with unfamiliar units. Some service members confront traumatic experiences or injuries during deployments, whereas others return unscathed. Some service members and their families feel more prepared or ready for an upcoming deployment than others. Almost all of these sources of variability in deployment characteristics lie outside of service members' control, but each could predict the nature of the aftermath of deployments for service members and their families.
- *Adaptive processes.* This construct refers to all of the ways in which family members interact, cope with stress, support each other, communicate, and resolve problems. Of particular interest are adaptive processes directly related to deployment, such as making efforts to communicate with the deployed service member, or moving to be closer to family while the service member is away. Adaptive processes, unlike deployment characteristics, are largely within the control of military families, and so should be facilitated or constrained by their preexisting conditions and should, in turn, directly account for immediate and long-term outcomes, such that more-effective adaptive processes should allow military families to endure or even thrive across the deployment cycle, whereas less effective adaptive processes should predict negative changes.

Immediate Outcomes

Some of the potential outcomes of the deployment experience may be evident immediately or soon after the deployment ends. We identified the following categories of immediate outcomes:

- *Emergent traits.* Family members may emerge from the experience of deployment irrevocably altered in ways that affect their subsequent functioning. Service members may return from a deployment with permanent physical or emotional injuries; they may return having grown or matured. In either case, they possess different capacities from those they had prior to the deployment. The family members who remain at home may be similarly altered in positive or negative ways (e.g., by additional education, noteworthy employment experiences, or new problems with substance abuse or the law).

- *Relationship quality.* The quality of the relationships among and between family members may be highly sensitive to the experience of deployment. Families in which these relationships were strong prior to the deployment and that adapt effectively during deployment should experience the least change across deployment and may even use the experience to draw closer. Families with problematic relationships prior to deployment and those without the skills or resources to adapt effectively during deployment may experience declines in the quality of their relationships across the deployment cycle. Either way, the postdeployment relationships will play a role in determining the long-term consequences for the family.

Long-Term Outcomes

The ultimate question that the Deployment Life Study is designed to address is how the experience of deployment affects the long-term well-being of military families. By closely examining those families that are successful, it is possible to start to paint a picture of what characteristics, qualities, and resources are related to family readiness. The dimensions of long-term well-being of greatest concern are *marital dissolution*, *military retention*, *child academic achievement*, *financial well-being*, *physical health*, and *mental, emotional, and behavioral health of all family members*.

All of these constructs can be assembled into the framework presented in Figure 3.1. The general shape of the figure is derived from the Karney and Crown (2007) model of military marriages, but it has been adapted to encompass the broader range of outcomes of interest in the Deployment Life Study.

The model suggests multiple pathways by which the experience of deployment may interact with preexisting conditions of military families to account for their immediate and long-

Figure 3.1
Conceptual Model of Military Family Health and Well-Being for the Deployment Life Study

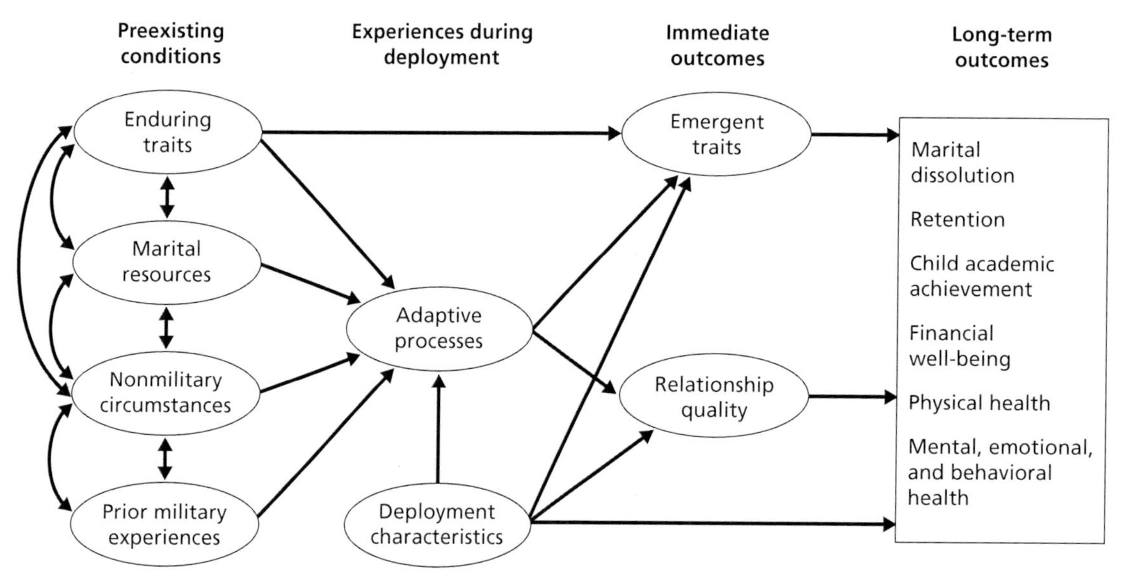

term outcomes after a deployment. First, the model suggests that all of the preexisting conditions of a family affect that family's outcomes through their direct effects on the quality of the family's adaptive processes during deployment. Adaptive processes, from this perspective, are the "skills and tools" that have been the defining features of family readiness. Placing these skills in a developmental context, the model describes the ability to adapt effectively as partially mediating the effects that a family's preexisting conditions can have on their postdeployment outcomes. For example, families should adapt more effectively to the extent that family members have enduring traits that provide strength, such as strong values, a commitment to marriage and the family unit, higher levels of education, or a history of mental health. Second, the model highlights the fact that the deployment itself is an exogenous variable that has direct effects on outcomes, as well as interactions with adaptive processes to affect a family's outcomes. The longer the deployment, for example, the greater demands the experience makes on a family's capacity to adapt effectively. Third, the model recognizes that military service can be a life-altering experience for each member of the family, leading to emergent traits that can be positive (e.g., personal growth) or negative (e.g., cognitive deficits associated resulting from TBI). This impact can be experienced at both individual and family levels because such events can disrupt how the family functions as a whole. Finally, the model recognizes that the consequences of deployment for families may evolve over time, such that the immediate impact can differ from the long-term impact.

The rest of the report describes how the Deployment Life Study baseline assessment was implemented, the measures used to assess the constructs in the conceptual model, the baseline sample, and how future, follow-up waves of data will be collected. The next chapter provides a discussion of study design, sampling and recruitment, and survey administration.

CHAPTER FOUR

Deployment Life Study Design

All methods, procedures, and instruments used in the study were approved by the RAND Human Subject Protection Committee. The survey instruments were licensed by the DoD Washington Headquarters Services in December 2010 (Record Control Schedule [RCS] HA [TRA] 2423). In addition, the study was granted a certificate of confidentiality from the National Institute of Mental Health (NIMH) (CC-MH-10-55).

Basic Study Design

The longitudinal study design used for the Deployment Life Study includes nine individual assessments with the service member, his or her spouse, and a child between the ages of 11 and 17 (if available) over a three-year period. In this manner, assessments are conducted every four months during this period (see Figure 4.1). A family is followed even if its service member retires or separates from the military or if the family experiences a divorce.

The unit of recruitment (and subsequent analysis) for the study is the household, defined as the service member, spouse, and child, if available.[1] Initial recruitment and the administration of the baseline assessment described in this report began in March 2011 and ended in August 2012 for the Army, Air Force, and Marine Corps sample and began in November

Figure 4.1
Deployment Life Study Timeline

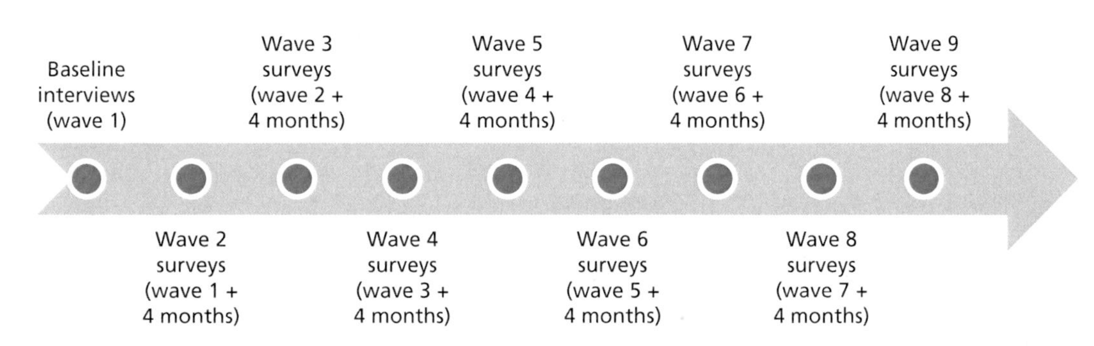

NOTE: Each dot indicates a four-month interval.
RAND *RR209-4.1*

[1] If an eligible child is not available or if an eligible child is available but does not participate in the survey, then the service member and spouse are considered a household.

2012 and ended in February 2013 for the Navy sample (see the next two sections for details on sampling and recruitment). Because not every household began the baseline survey at the same time, the enrollment period was considered rolling, with subsequent surveys anchored to a household's baseline completion. For example, if a household enrolled in March 2011, the first follow-up survey was in July 2011 (i.e., four months after baseline). However, if a household enrolled in December 2011, its first follow-up was in April 2012.

The longitudinal design was constructed to allow for the collection of baseline data and, on average, one to three additional follow-up assessments prior to deployment, approximately one to three assessments during deployment (depending on length of deployment), and approximately one to four assessments postdeployment (depending on timing of redeployment, or when the service member returns home). Because deployment characteristics vary by service, the number of survey waves completed before, during, and after deployment may differ across families.

Sampling

To identify a group of military families eligible for participation in the Deployment Life Study, each of the services (i.e., Army, Navy, Air Force, and Marine Corps) provided a list of deployment-eligible service members. A request was submitted to a point of contact (POC) within each service to provide administrative data on approximately 22,000 service members from both the active and reserve components. The office designation for each of the service-specific POCs varied depending on which entity within a service had access to personnel data and had content expertise and interest in the study. For example, the study team partnered with the office of the Army Surgeon General, U.S. Army Human Resources Command (HRC), U.S. Army Reserve (USAR) Military Personnel Management (G-1), and the National Guard Bureau (NGB) to obtain data for the Army sample; for the Air Force, the study team worked with leaders focused on family readiness issues (Airman and Family Readiness Policy [A1SA]); for the Marine Corps, we worked with the Personal and Family Readiness Division within Manpower and Reserve Affairs (M&RA); and, for the Navy, the Navy Bureau of Medicine and Surgery (BUMED) (medical command) and the Bureau of Naval Personnel (BUPERS) (personnel data). The requested data for each service member included name, gender, marital status, rank, pay grade, occupational code, unit identifier code, address, telephone number, email address, and number of prior deployments.

Once this information was obtained from the services, construction of a recruitment pool began, with the aim of a sample size of 16,000 service members who would be drawn randomly from the 22,000 in the target sampling frame. Service members eligible to participate included those who were married and living with their spouses and were deemed by their respective services to be deployable (i.e., eligible to deploy) six months to one year following the baseline interview. Once deployment information was obtained, the intent was to sample components (active and reserve) in numbers proportional to their actual representation among deploying married service members, oversampling married female service members and aiming for a minimum of 500 families for the active component in each service and a minimum of 150 families from each reserve component. Thus, the original design called for a sample-stratified design on service, component, and gender.

Sample data were initially obtained in 2010 from the Army, Air Force, and Marine Corps. Because of delays in DoD-required approvals for data collection and thus baseline fielding, refreshed and updated sample files for all services were obtained in early 2011. The files received from the Army contained more individuals than expected, whereas the sampling files from the Air Force and Marine Corps contained far fewer eligible individuals than originally anticipated. These discrepancies were likely a function of the currently married eligibility criteria and the changing nature of the deployment rotations scheduled for 2011 (as compared with the higher numbers being deployed in 2009 and 2010). We received data from the Navy in September 2012. The Defense Manpower Data Center and Navy leadership (specifically, BUPERS) linked personnel records with information on which units were deployable. From this list, we were able to sample households with a sailor eligible to be deployed in the next 12 months.

Table 4.1 details the initial study sampling frame (all households provided) and final number of respondents by service and respondent type. The Army provided 54,993 eligible names; the Air Force provided 8,273; and the Marine Corps provided 2,710 (see second column). Initially, the intention was to capture a large enough sample across Army, Air Force, and Marine Corps (n = 65,976) at baseline so that, after accounting for attrition across the waves (see third column), the final sample at wave 9 would total roughly 4,000 households for those three services (see fourth column). However, because of quality issues with the phone numbers (missing or incomplete numbers), sample expectations were modified during baseline enrollment, with a goal of achieving roughly 1,650 baseline households across the three services (see fifth column).

Because the feasibility of the Navy sample was still in doubt during the baseline phase for the Army, Air Force, and Marine Corps sample, we did not adjust our targeted sample size once it was clear that we would be able to recruit Navy families. The Navy provided us with a total of 28,529 eligible names, more than the requested 22,000. Sample targets are provided in Table 4.1. Note that we did not readjust our target because the quality of contact data was higher in the Navy sample than in the Army, Air Force, and Marine Corps sample.

The initial sampling design included a plan to use multiple gender-stratified random samples within each component (a process in which the different samples are commonly referred to as *sample replicates*). The plan was to slowly release batches of gender-stratified random samples within each component in such a way that later sample replicates would be used only if response rates within a given stratum did not yield the targeted sample size within that stratum. Originally, the plan called for using two replicates for each stratum; however, unexpected challenges with respect to the quality of the sampling data provided by DoD led to much lower qualification and response rates than initially expected. This required using more than two additional replicates in the Army and Air Force and use of the full available sample data provided from the Marine Corps. The additional replicates in the Army and the Air Force were no longer required to stratify by gender once it became readily apparent that the targeted sample size for female married and deployable service members within these services was not going to be properly met. In the Air Force Reserve, gender stratification could not be implemented because gender was not available in the sampling file. There were no replicates in the Navy file; all numbers were eligible and released for screening calls.

Although the random samples of service members within each component were representative of the populations of service members from which they were sampled, the sample of respondents with baseline surveys within each component did differ systematically from the

Table 4.1
Initial Deployment Life Study Sampling Frame

Service and Component	Households from Service[a]	Deployment Life Study Target Household Baseline Sample[b]	Projected Target Household Sample by End of Study Period[c]	New Household Baseline Sample Projection (based on n = 1,651)[d]	Actual Respondents[e]		
					Service Member	Spouse	Child
Air Force, Army, Marine Corps sample							
Air Force	8,273	1,845	1,125	460	420	317	48
Active	6,913	1,330	811	332	351	273	35
Guard	346	269	164	67	48	32	10
Reserve	1,014	246	150	61	21	12	3
Army	54,993	3,696	2,253	925	1,942	1,492	343
Active	38,365	2,610	1,591	654	1,034	791	157
Guard	15,190	781	476	195	824	640	176
Reserve	1,438	305	186	76	84	61	10
Marine Corps	2,710	1,066	650	266	187	143	11
Active	2,437	820	500	205	161	123	10
Reserve	273	246	150	61	26	20	1
Total	65,976	6,607	4,028	1,651	2,549	1,952	402
Navy sample							
Navy							
Active	28,085	1,348	822	1,348	1,036	893	103
Reserve	444	246	150	246	10	8	3
Total	28,529	1,594	972	1,594	1,046	901	106
Total	94,505	8,201	5,000	3,245	3,595	2,853	508

Table 4.1—Continued

[a] Cells represent service member data provided to RAND by the Air Force, Army, Navy, and Marine Corps. The sample reflects all names provided by the services, regardless of phone quality.

[b] Cells represent data projections for initial baseline sample. As described earlier, these expectations were adjusted to n = 1,651 for Army, Air Force, and Marine Corps because of the phone number–quality concerns. No adjustments were made for the Navy sample. Households are defined as a service member and spouse. Households do not necessarily contain a child.

[c] Cells represent goals for household totals at the final wave (assuming attrition over the three study years).

[d] Baseline sample projections based on new total households (n = 1,651). This does not apply to the Navy sample.

[e] These data include all completed surveys, regardless of whether the household was enrolled. For example, a service member may have completed the baseline survey (and is included in the service member column) even if his or her spouse may not have. This is why the columns do not sum to the final household number of 2,724 and why the data in the columns for service members and spouses are not the same. These data also include surveys completed both via phone and on the web.

overall population of service members in the sampling files. To compensate for the nonrandom nature of the final sample, analytic weights were created to ensure that the characteristics of the baseline sample of respondents reflect the characteristics of the individuals in the sampling frame files provided by DoD (see "Analytic Weights" later in this chapter for more details on the construction of sampling weights).

Given the lower-than-expected final sample size, power calculations were reexamined in order to explore the implications for the estimate precision. Although the limits on sample size will preclude some within-service analyses (e.g., by component within the Marine Corps), most analyses comparing outcomes across services will proceed as originally planned.

Recruitment

Once the sampling frame was identified, the sampling frame lists, with contact information, from each service and component were given to a subcontractor, Abt SRBI. Abt SRBI is a survey company with experience in military family studies. Between March 2011 and June 2012 and again between November 2012 and February 2013, Abt SRBI recruited service members and their families, in a random fashion, from those lists. To initiate this effort, Abt SRBI sent selected service members an introductory letter from the principal investigators (PIs) of the Deployment Life Study. These study recruitment materials included a prenotification letter describing the study, the study funders, and basic eligibility criteria and gave assurance that participation was voluntary and confidential. The letter informed the service member and his or her family that an interviewer would try to reach the household by phone in the next few days, or, if the service member preferred, he or she could call in to Abt SRBI and provide a unique personal identification number (PIN) provided in the letter in order to schedule his or her interview. The mailing also included an endorsement letter from an official from the service that the potential respondent represented. Shortly after recruitment started, personalized recruitment email messages were sent to sample service members with an email address. The email contained the same information as the recruitment letter and included a link where the service member could update his or her contact information, provide an updated telephone number, and, in some cases, provide suggested times to call for an interview.

Abt SRBI then began calling the landlines and cell phones of service members in the sample lists. In order to maximize enrollment, Abt SRBI made efforts to recruit as many households as possible, including the following techniques:

- setting a maximum number of attempts to contact households after which no further calls were made. For households in which at least one interview was completed, the maximum number of attempts to reach additional respondents was set at 50. For phone numbers for which no interview had been completed (e.g., no answers, busy signals, calls in which the person answering could not complete the screening and interview at the time of the call), the maximum number of attempts was set at 15.
- discontinuing attempts to complete an interview in certain circumstances, such as reaching a number that was disconnected or not working, getting a wrong number, reaching a respondent who has health problems, reaching a Spanish-speaking-only respondent, or being told that the person the interviewer was attempting to contact is deceased or ineligible

- making multiple calls until partial or complete interviews were completed for up to three eligible household respondents
- interviewing all respondents during a single call in which the phone is passed from one respondent to another when possible, minimizing the need for follow-up calls to the household to reach additional respondents.

To aid in recruitment, potential survey participants also received additional letters and emails if they fell into one of the following categories in the Abt SRBI recruitment efforts: noncontacts, time-outs, and web-only baseline:

- *Noncontacts:* After the initial attempts to reach a household, if Abt SRBI had a working phone number or, in many cases, had an initial conversation with either the service member or a spouse but was unable complete an interview, a letter and email message were sent encouraging the service member to enroll. That household was then put back in the queue for Abt SRBI to resume calling.
- *Time-outs:* When Abt SRBI successfully completed the first adult's baseline interview and four weeks or more had gone by without successfully completing the second adult interview, we sent a letter and email reminding the adult who had not yet completed the interview that he or she had not yet done so. This correspondence emphasized that the household could not be enrolled in the study unless both adults completed the baseline interview. Near the end of the recruitment period, if Abt SRBI was still unable to complete an interview (and thus complete a household) and had exhausted its attempts to contact the respondent, we sent a final recruitment email to these households and invited them to complete the baseline survey on the web. Approximately 13,500 potential participants received this email.
- *Web-only baseline:* In August 2011, an email recruitment procedure and online administration of the baseline interview were used with a small sample. This pilot attempted to contact 400 potential participants who had email addresses but no phone numbers. Two hundred of these participants were sent a recruitment email and directed to an online version of the baseline interview. Two weeks later, the remaining 200 received the same email. Ultimately, we continued this approach for those households that did not have a mailing address or phone number. If a service member completed the survey online, he or she was instructed on how to provide information so that his or her spouse and (if applicable) child could access their respective surveys.

Use of Study Incentives

The success of any longitudinal research study depends heavily on continued participation of the research sample over time. Participants in the Deployment Life Study are asked for a significant commitment over nine waves of data collection. Prior research experience suggests that one powerful way to ensure continued participation over a longitudinal study is to acknowledge the participants with tokens of appreciation (see Laurie and Lynn, 2009). At the time of our study approval, the Defense Manpower Data Center (DMDC) would not permit the provision of direct monetary incentives to service members; instead, tokens of appreciation are provided to the household in the form of gift cards to spouses and available study children.

Over the course of the entire study, families are eligible for up to $655 in Visa gift cards. These tokens of appreciation are administered in payments after each wave of data collection.

For the baseline, spouses who completed the interview received a Visa gift card valued at $50 whether or not the household ultimately enrolled in the study. Children who completed the baseline measure received a $25 gift card. During follow-up surveys, a household (in the name of the spouse) receives $40 for each wave of data for which one adult (i.e., either the service member or spouse) completes the assessment, with an additional $20 card for each assessment the enrolled child completes. Families will also receive a bonus of $100 if at least ten out of 18 adult surveys (spouse + service member × nine waves) are completed at the conclusion of the study.

Screening and Consent of Adult Participants

Because inclusion in the sample was dependent on the service member being married, having the potential to be deployed in the near future, and intending to remain in the military, a fairly lengthy process screened possible participants before the baseline survey could be completed (see Appendix B, online at www.rand.org/pubs/research_reports/RR209, for a copy of the screening instrument). Upon reaching either the service member or spouse by phone, interviewers began screening potential participants for eligibility.[2] Interviewers reminded potential participants of the purpose of the study and referenced the recruitment mailing that participants should have received. They asked questions regarding the following:

- marital status
- living arrangements with spouse
- intentions to leave the service soon
- deployment status of the household service member.[3]

Households were deemed eligible to enroll if spouses were married and living together and the service member did not plan to leave service within the next six months (i.e., did not file separation or retirement paperwork and had filed reenlistment or extension paperwork if within six months of his or her enlistment period).

In addition to the screening requirements, two other requirements existed for household eligibility. First, both the service member and the spouse had to agree to participate in the study and provide oral consent to participation. Second, both service member and spouse (and child, if available) had to complete the baseline interviews within eight weeks of the first survey completion by a household member (i.e., if the spouse completed first, the service member [and child] had to complete their surveys within the following eight weeks). The administration window was required because the study intended to capture each respondent's perspectives on a comparable period in the family experience. It was therefore possible, regardless of which adult was interviewed first, that the household would fail to qualify if the second adult did not also complete the interview in the required time frame.[4]

---

[2] If a spouse answered the phone, the interviewer proceeded with a screening script that determined household eligibility and could segue directly into an interview.

[3] Eventually, the screener stopped asking service members whether they were aware of an upcoming deployment because too many respondents answered in the negative and were thus screened out. Recall that the services provided a list of service members who were scheduled to deploy within a year of the baseline.

[4] If the service member in the household was currently deployed, his or her spouse could complete the baseline interview if the service member was expected to return within eight weeks. If the service member was expected to be deployed for longer than eight weeks, the spouse was not enrolled, and their household was placed on a callback list for later contact.

Screening, Consent, and Assent for Child Participants

After at least one adult consented to enroll in the study, the interviewer asked a series of questions in an attempt to identify and recruit an eligible child for the study. Participants were asked to report whether any children under the age of 18 lived in the household. Because only children between the ages of 11 and 17 were eligible to participate as the *study child*, the interviewer determined whether there were children in that age range and, if there were more than one, selected the child whose first name came first alphabetically. Once a child was selected, the interviewer confirmed the relationship of the service member and spouse to the child (e.g., parent, stepparent, foster parent) and the residential status of the child (i.e., whether the child lived in the household at least half the time during the past six months). If an eligible child was identified, the responding adult was asked to consent for his or her child to be part of the study. Children were also required to give their assent to participate before being enrolled.

Note that, if the family did not have an eligible child to be a study child, the service member and his or her spouse were still queried about children in the household as part of their surveys. For a family in that situation, a child between the ages of three and 18 years was randomly selected to serve as the *focal child*. Adults were asked about that child's health and well-being, and that focal child was the subject of questions from baseline through the final survey wave.

Sampling Flow

Army, Air Force, and Marine Corps Sample

Table 4.2 outlines the sampling flow for the Army, Air Force, and Marine Corps sample, starting with the total number of service members for whom eligibility information (e.g., basic demographic data, such as marital status) and contact information were obtained, through the number of households that completed the screener and were eligible to participate in the baseline survey. Please note that the sample was reduced from the numbers shown in Table 4.1 because of phone-number and email-address availability in the records. Of the roughly 20,000 numbers dialed, only 64 percent were deemed "good." In this context, a good number is one that is in working order (i.e., not disconnected). This means that 36 percent of the numbers provided to Abt SRBI were not valid.

Roughly 93 percent of all good numbers led to contact with an actual person. Of those, about 6,400, or 54 percent, were not screened. Several factors led to an unscreened, but good, number: answering machines, call blocking, callbacks that were never completed, and hard refusals. Of note is that 34 percent of those numbers that were contacted but not screened refused to participate. From a total of 20,138 dialed numbers, 5,554 (28 percent) were screened for eligibility to participate in the Deployment Life Study.

Table 4.2 also shows the 21,220 cases included in our web-only sample. This sample included service members who did not have valid phone numbers (the field may have been blank or incomplete in the data files) but did have valid email addresses. Of those, 764 households were at least partially screened, with the vast majority (more than 90 percent) completing the full screen. There are several reasons that the number and proportion screened are so much lower than the number of cases in our web-only sample. It is possible that email addresses were inaccurate (i.e., "bounced back"), that emails were routed to spam or junk-mail folders, or that service members simply did not want to participate in the survey.

Table 4.3 presents the disposition of the 5,554 screened calls for the Army, Air Force, and Marine Corps sample. Roughly 59 percent were screened and deemed ineligible. The majority

Table 4.2
Sample Flow of the Deployment Life Study Through the Screener: Army, Air Force, and Marine Corps Sample

Status	Number	Percentage
Total numbers dialed by Abt SRBI	20,138[a]	—
Total good numbers	12,818	64% of dialed numbers
Total contacts	11,961	93% of eligible (good) numbers
Not screened	6,407	54% of contacted numbers
Dead number[b]	365	6% of not screened
Callback not completed	2,745	43% of not screened
Answering machine or call blocked	1,115	17% of not screened
Refused screening	2,182	34% of not screened
Individuals screened by phone	5,554	46% of contacted numbers
Web-only sample		
Email contact–only sample[c]	21,220	—
Households screened by web	764	
Completed screener	694	91% of screened
Partially completed screener	70	9% of screened

[a] The total sample dialed reflects only those households for which we had valid phone numbers.

[b] Examples of dead numbers are those for which the respondent who answered the phone spoke primarily a foreign language, was a minor with no adult present, or had a health or hearing problem that affected his or her ability to participate.

[c] This sample contained service members for whom valid phone numbers were not available but for whom an email address was present.

of these "screen-outs" occurred because the service member was retired or in the process of separating from the military, the service member and spouse were divorced or not living together, or the service member was deployed at the time of the call. Less than 10 percent of screened numbers timed out because a household completed a screener but could not be contacted to complete the baseline survey. Roughly 5 percent of the screened and eligible numbers declined to participate in the baseline survey. The result was 1,592 households completed by phone, via Abt SRBI. The rest of the baseline surveys were completed (n = 259) via web.

To be considered a completed household, a household had to have, at minimum, a completed survey by both the service member and the spouse. As indicated in Table 4.3, among phone completions, 1,238 households included surveys from only a service member and spouse. An additional 354 households also included a study child. The table also shows the disposition for the web-only screened sample. Of the 259 households enrolled via the web, 229 households made up of a service member and spouse completed the baseline survey, with an additional 30 households made up of a service member, spouse, and child.[5]

[5] The discrepancy between the number of households screened via the web (n = 764; see Table 4.2) and the number of households enrolled via the web (n = 259; see Table 4.3) is due to ineligibility (e.g., the service member being retired or retir-

Table 4.3
Disposition of Screened Calls for the Army, Air Force, and Marine Corps Sample

Status	Number	Percentage
Total phone numbers screened by Abt SRBI	5,554	—
Ineligible (screen-outs)[a]	3,251	59% of screened numbers
Dead callbacks[b]	407	7% of screened numbers
Refused	272	5% of screened numbers
Interviews completed by Abt SRBI	1,593	29% of screened numbers
Baseline Deployment Life Study sample		
Phone		
Households[c]	1,592	—
Service member and spouse	1,238	—
Service member, spouse, and child	354	—
Web only		
Households[c]	259	—
Service member and spouse	229	—
Service member, spouse, and child	30	—
Total sample	1,851	
Service member and spouse	1,467	
Service member, spouse, and child	384	

[a] Examples of screen-outs include households in which the service member has retired or separating from the military, the service member and spouse were divorced or not living together, or the service member was deployed at the time of the call.

[b] A dead callback is a screened, eligible case that delayed completion of the baseline survey but Abt SRBI was not able to reestablish contact with the household. This differs from "callback not completed" in Table 4.2, in which no one in the household was actually reached and thus the household is never screened for eligibility.

[c] A completed household is defined as one in which both the service member and spouse completed the baseline survey, regardless of the completion status of an eligible child. One service member self-identified as being in the Navy. That case is included in Tables 4.4 and 4.5 but not here.

Navy Sample

Table 4.4 outlines the sampling flow for the Navy sample, starting with the total number of service members for whom eligibility information (e.g., basic demographic data, such as marital status) and contact information were obtained, through the number of households that completed the screener and were eligible to participate in the baseline survey. Please note that the sample was reduced from the numbers shown in Table 4.1 because of phone-number and email-address availability in the records. Of the roughly 15,000 numbers dialed, 80 percent were deemed "good." In this context, a good number is one that is in working order (i.e., not

---

ing, the service member being deployed at the time of the baseline survey, or the service member not living with his or her spouse).

Table 4.4
Sample Flow of the Deployment Life Study Through the Screener: Navy Sample

Status	Number	Percentage
Total numbers dialed by Abt SRBI	14,946[a]	—
Total good numbers	11,970	80% of dialed numbers
Total contacts	11,039	92% of eligible (good) numbers
Not screened	8,589	78% of contacted numbers
Dead number[b]	129	2% of not screened
Callback not completed	4,350	51% of not screened
Answering machine or call blocked	2,847	33% of not screened
Refused screening	1,263	15% of not screened
Individuals screened by phone	2,450	22% of contacted numbers
Web-only sample		
Email contact–only sample[c]	2,343	—
Households screened by web	65	
Completed screener	58	89% of screened
Partially completed screener	7	11% of screened

[a] The total sample dialed reflects only those households for which we had valid phone numbers.

[b] Examples of dead numbers are those for which the respondent who answered the phone spoke primarily a foreign language, was a minor with no adult present, or had a health or hearing problem that affected his or her ability to participate.

[c] This sample contained service members for whom valid phone numbers were not available but for whom an email address was present.

disconnected). This means that 20 percent of the numbers provided to Abt SRBI were not valid.

Roughly 92 percent of all good numbers led to contact with an actual person. Of those, about 8,600, or 78 percent, were not screened. Several factors led to an unscreened, but good, number: answering machines, call blocking, callbacks that were never completed, and hard refusals. Of note is that 15 percent of those numbers that were contacted but not screened refused to participate. From a total of 14,946 dialed numbers, 2,450 (22 percent) were screened for eligibility to participate in the Deployment Life Study.

Table 4.4 also shows the 2,343 cases included in our web-only sample. This sample included service members who did not have valid phone numbers (the field may have been blank or incomplete in the data files) but did have valid email addresses. Notice that there were far fewer of these cases in the Navy sample than in the Army, Air Force, and Marine Corps sample; this occurred because the quality of the telephone data was higher in the Navy data. Of those, 65 households were at least partially screened, with the vast majority (almost 90 percent) completing the full screen. There are several reasons that the number and proportion screened are so much lower than the number of cases in our web-only sample. It is possible that email addresses were inaccurate (i.e., "bounced back"), that emails were routed to spam or junk-mail folders, or that service members simply did not want to participate in the survey.

Table 4.5 presents the disposition of the 2,450 screened calls for Navy sample. Roughly 47 percent were screened and deemed ineligible. The majority of these "screen-outs" occurred because the service member was retired or in the process of separating from the military, the service member and spouse were divorced or not living together, or the service member was deployed at the time of the call. Roughly 15 percent of screened numbers timed out because a household completed a screener but could not be contacted to complete the baseline survey. Roughly 4 percent of the screened and eligible numbers declined to participate in the baseline survey. The result was 862 households completed by phone, via Abt SRBI. The rest of the baseline surveys were completed (n = 11) via web.

To be considered a completed household, a household had to have, at minimum, a completed survey by both the service member and the spouse. As indicated in Table 4.5, among

Table 4.5
Disposition of Screened Calls for the Navy Sample

Status	Number	Percentage
Total phone numbers screened by Abt SRBI	2,450	—
Ineligible (screen-outs)[a]	1,141	47% of screened numbers
Dead callbacks[b]	360	15% of screened numbers
Refused	88	4% of screened numbers
Interviews completed by Abt SRBI	862	35% of screened numbers
Baseline Deployment Life Study sample		
Phone		
Households[c]	863	—
Service member and spouse	759	—
Service member, spouse, and child	103	—
Web only		
Households[c]	11	—
Service member and spouse	10	—
Service member, spouse, and child	1	—
Total sample	873	
Service member and spouse	768	
Service member, spouse, and child	104	

[a] Examples of screen-outs include households in which the service member was retired or separating from the military, the service member and spouse were divorced or not living together, or the service member was deployed at the time of the call.

[b] A dead callback is a screened, eligible case that delayed completion of the baseline survey but Abt SRBI was not able to reestablish contact with the household. This differs from "callback not completed" in Table 4.3, in which no one in the household was actually reached and thus the household is never screened for eligibility.

[c] A completed household is defined as one in which both the service member and spouse completed the baseline survey, regardless of the completion status of an eligible child. One Navy service member completed the survey as part of the Army, Air Force, and Marine Corps sample.

phone completions, 759 households included surveys from only a service member and spouse. An additional 103 households also included a study child. The table also shows the disposition for the web-only screened sample. Of the 11 households enrolled via the web, ten households made up of a service member and spouse completed the baseline survey, with one additional household made up of a service member, spouse, and child.[6]

The total baseline sample is made up of 2,724 households: 2,236 in which a service member and spouse completed surveys and 488 in which a service member, spouse, and child completed surveys. Table 4.6 presents the final baseline sample, broken down by service and component. Please note that the data in this table are *unweighted*; weighted percentages of the sample by service and component are presented in Table 6.3 in Chapter Six.

Table 4.6
Final Deployment Life Study Baseline Sample, by Service and Component, Raw Ns and Unweighted Percentages

Service and Component	Raw N	Unweighted Percentage
Air Force	298	10.9
Active	254	9.3
Guard	32	1.2
Reserve	12	0.4
Army	1,426	52.3
Active	773	28.4
Guard	591	21.7
Reserve	62	2.3
Navy	873	32.0
Active	866	31.8
Reserve	7	0.3
Marine Corps	127	4.7
Active	110	4.0
Reserve	17	0.6

NOTE: n = 2,724. Families are allocated to service and component according to service member self-report data in the baseline survey.

[6] The discrepancy between the number of households screened via the web (n = 65; see Table 4.4) and the number of households enrolled via the web (n = 11; see Table 4.5) is due to ineligibility (e.g., the service member being retired or retiring, the service member being deployed at the time of the baseline survey, or the service member not living with his or her spouse).

Analytic Weights

Analytic weights were created to minimize the effects of nonresponse bias among the baseline sample of completed households. Specifically, weights can be used to make the study sample at baseline representative of married, deployable service members in the field around 2011 and 2013 when this study was launched. Computing the weights included three key steps:

1. Create weights, which made the baseline sample within a given service and component look like the data file from which it was sampled (which was a random sample of married service members who were likely to be deployed within six to 12 months of receipt of the files from the given service and component in question).
2. Adjust weights within a service's sample to ensure that the proportion of service members across components in our study was representative of married, deployed service members within that service. This adjustment was made based on the observed proportion of service members within a service's components using personnel data on married service members who were deployed some time between June 1, 2012, and December 31, 2012.
3. Adjust weights across all services to ensure that the proportions of service members across services in our study were representative of married service members who were deployed some time between June 1, 2012, and December 31, 2012. This was based on the same personnel data as was used for step 2.

Table 4.7 shows a comparison of the unweighted and weighted households in the study for each service. Before applying the weights, we see that a higher percentage of officer-headed households completed the survey than in the sampling file that we received from each service. For example, 30 percent of service members in Army households in the baseline sample were officers, versus 15 percent in the sampling file received from the Army. For the Air Force, 31 percent of completed households were those of officers, versus 24 percent in the sampling file. For the Marine Corps, it was 58 percent versus 45 percent, and, for the Navy, it was 20 percent versus 11 percent. This may be due to the marital-status eligibility criteria. As also shown in Table 4.7, the analytic weights corrected each of these key imbalances and improved other, more-moderate imbalances observed between the completed households and the universe from which they were sampled.[7]

Survey Administration

Pretest

The survey instruments were pilot-tested before fielding in two stages. First, the service member and spouse instruments were fielded among a sample of military personnel placed at RAND under a military fellowship. These individuals were not part of the study team and mirrored many of the characteristics of our proposed Deployment Life Study sample. The fellows (n = 5)

[7] Analytic weights are focused on addressing nonresponse on observable characteristics. The pronounced differences seen in Table 4.7 are not seen when comparing the random samples of service members for recruitment into the study with the sampling file. Systematic differences occur only when the characteristics of households that completed the baseline survey are compared with those in the original sampling files.

Table 4.7
Comparing Characteristics of Unweighted and Weighted Household Sample with the Original Sampling File for Each Service

	Sampling File		Completed Households	
Characteristic	Mean	Standard Deviation	Unweighted Mean	Weighted Mean
Air Force				
Male	0.87	0.34	0.84	0.89
Officer	0.23	0.42	0.31*	0.23
Number of previous deployments	2.45	2.32	2.36	2.31
Army				
Male	0.91	0.29	0.94	0.92
Officer	0.15	0.35	0.30*	0.16
Number of previous deployments	1.43	1.15	1.61	1.53
Number of dependents	2.10	1.54	1.80	2.18
Marine Corps[a]				
Officer	0.45	0.50	0.58*	0.48
Number of previous deployments	1.86	1.42	2.06	2.04
Navy				
Male	0.91	0.29	0.93	0.91
Officer	0.11	0.31	0.20*	0.11
Number of previous deployments	1.88	1.22	2.09	1.87

* Denotes when the standardized mean difference between our completed households and the sampling file population was greater than 0.20.

[a] More than half of the information on gender was missing in the Marine Corps sample file and thus was not used in the creation of the weights. However, the Marine Corps is predominantly male.

were asked to review the survey for content, question meaning and interpretation, and ease and length of survey. Each completed the instruments for timing as well. As a result of their feedback, several changes were made to the instruments, including changes to the order of questions, overall instrument length (the survey was shortened), and specific question wording for clarity. In addition to this pretesting, the first 20 interviews completed during the fielding period were used as a pilot test for timing and length, interpretation, and readability. Using the feedback from these 20 cases, we made further revisions for clarity and streamlining before the official start for the baseline field period.

Baseline: Wave 1

As noted above, rolling enrollment meant that baseline service member, spouse, and child assessments took place between March 31, 2011, and August 31, 2012, via computer-assisted telephone interview (CATI), with a limited number of potential participants completing the baseline survey via the Internet. Not including the initial screening and consent, the baseline interviews lasted approximately 48 minutes for each adult and 26 minutes for each child. Baseline survey instruments can be found in Appendix C (for service members), Appendix D (for spouses), and Appendix E (for children), online at www.rand.org/pubs/research_reports/RR209.[8]

Follow-Up Surveys: Waves 2 Through 9

Households will receive invitations to complete follow-up surveys (i.e., waves 2 through 9) via email, as well as postal mail, inviting them to participate in the next wave of the study, approximately four months after the previous wave. Similar to the baseline, follow-up surveys remain open for eight weeks. However, the eight-week "clock" begins not when the first household member completes his or her survey but rather on the date that the household is eligible to complete the follow-up (i.e., four months after the household completed the baseline interview and thus was "officially" enrolled).

According to this timing system, participants who complete the survey near the end of the eight-week window still get the next invitation four months from the previous invitation. Thus, it is possible that a household could complete consecutive waves anywhere between two months and six months after the previous wave.

Follow-up assessments are being conducted primarily as an online survey over a secure website, with an option for completing the survey via telephone with a live interviewer. Estimates suggest that between 75 and 90 percent of the military families in our baseline sample will have access to a computer (public or private) and the Internet and will choose to complete follow-up surveys via this medium. The remaining participants will request or require that their surveys be conducted over the telephone. The RAND Survey Research Group (SRG) will complete the follow-up interviews. SRG will provide reminder phone calls to households that are at risk of not completing during the eight-week window. During these calls, which occur four weeks after the first household member completes his or her survey, SRG will offer the participant the opportunity to complete the survey via telephone.

The following guidelines are used for ongoing participation in the Deployment Life Study:

- Even if a respondent—or all the participating family members—misses or skips a survey, the household is still considered enrolled unless it requests to be removed from the study.
- If one adult wants to drop out but the second adult would like to remain enrolled, each will be allowed to take his or her desired action.
- A child cannot remain in the study unless at least one adult respondent in his or her household remains enrolled.
- Once a household is enrolled, if the marriage dissolves, both adults can continue to participate in the study and complete truncated versions of the instruments.

Of note is the fact that, in each follow-up interview, each respondent will be asked whether the service member is currently preparing for deployment, currently deployed, or recently returned from deployment. A module of survey questions specific to that aspect of deployment will be added to that survey wave for each family member. To encourage retention of the sample, updated contact information is collected at each subsequent wave of data collection.

[8] A fuller discussion of survey items can be found in Chapter Five of this report.

The next chapter of the report shifts to a discussion of the measures and survey items used to assess family readiness and the impact of deployment on family health and well-being. The discussion includes a description of key constructs and measures in the Deployment Life Study, using the conceptual model described in Chapter Three (see Figure 3.1) as a guide.

CHAPTER FIVE

Constructs and Measures

This chapter describes, in more detail, the rationale for specific constructs within the four main domains assessed in this study (i.e., preexisting conditions, experiences during deployment, immediate outcomes, and long-term outcomes). It also describes the specific variables used to operationalize each construct.

It is important to keep in mind that many variables can be treated as measures of more than one construct, depending on the research question. For example, a service member's mental health could be considered a preexisting condition (i.e., an enduring trait) when used to predict marital dissolution. Yet, in another analysis, service member mental health may be a long-term outcome predicted by deployment experiences.

Not every measure included in the Deployment Life Study assessments is described in detail in this chapter. The focus here is on a selected group of measures that exemplify each construct.[1] In general, the research team relied on existing measures that have been widely used across diverse samples (e.g., civilians and military, racially and ethnically diverse, international, low income) and validated in the existing literature. Although it is outside the scope of this report to present the psychometric properties of each of the measures across different samples, references are provided so that the interested reader can find such information.

Throughout this report, the term *study child* refers to a child who participates in the Deployment Life Study by completing his or her own survey. A *focal child* is a child for whom the service member and spouse are asked questions but who does not him- or herself complete the child survey.

Preexisting Conditions

Enduring Traits

Enduring traits of family members are considered preexisting conditions that may modify, either by mitigating or exacerbating, the association between military deployment and family health well-being. Grouped within this construct are all of the relatively stable qualities of family members that prior research on resilience and vulnerability suggests will affect family members' readiness for deployment. Specific constructs measured within this theme include the following:

[1] The complete set of items in the baseline assessments for service members, spouses, and children can be found in Appendixes C, D, and E, respectively, online at www.rand.org/pubs/research_reports/RR209.

- *Sociodemographic characteristics.* At baseline, each participant reported his or her gender, age, race, ethnicity, educational attainment, and employment status. Characteristics that are stable over time (e.g., gender, race, ethnicity) will not be measured again, whereas characteristics that may change (e.g., educational attainment, employment status) are measured at every assessment and may be treated as outcomes and as preexisting conditions.
- *Family of origin.* Characteristics of service members' and spouses' families of origin may function as sources of vulnerability. For example, individuals whose parents divorced or were never married are less likely to have stable marriages than individuals whose parents were married and remained married (Glenn and Kramer, 1987). Families of origin may also serve as sources of resilience. For example, individuals from military families may have more-realistic expectations regarding deployment than individuals who were not raised within the military culture. Service members and spouses were asked (1) whether their parents were married when the respondent was born; (2) if their parents were married, whether, how (e.g., divorce or death), and when the marriage ended; and (3) whether their parents had served in the military. Because these variables do not change over time, these items were administered at baseline only.
- *Adverse childhood experiences.* The personal history of each family member is another characteristic that does not change over time. Of the many aspects of personal history that could have been studied, the Deployment Life Study administered a subset of items from the Adverse Childhood Experiences (ACE) scale (Felitti et al., 1998; Anda et al., 1999). Among these are items asking whether respondents experienced abuse or neglect in childhood (e.g., feeling physically threatened, did not have enough to eat, had a parent who went to prison). Higher scores on the ACE have been associated with negative outcomes (e.g., depression, substance abuse, coronary artery disease) in adulthood (Felitti and Anda, 2010), so we expect that individuals endorsing greater numbers of these adverse experiences will have a more difficult time coping with the stress of deployment as well. These items were administered to service members and spouses at baseline only.
- *Spirituality and religiosity.* In prior research on civilian families, those who report greater feelings of spirituality and closer connections to religious organizations cope with stress more effectively than people in families without those connections (e.g., Park, Cohen, and Herb, 1990). Both of these constructs are likely to be relatively stable over time, but military families may also embrace religion and religious experience as coping mechanisms during and after deployments (e.g., Ano and Vasconcelles, 2005). To measure these constructs, the baseline assessment included four items that asked respondents to report on the religion with which they identify most, the frequency with which they attend religious services, and the degree to which spiritual or religious beliefs influence the way they live and cope with problems or difficulties. These items are administered again in all subsequent survey waves. These items have been used in the Florida Project on Newlywed Marriage and Adult Development (FPNMAD) (Trail and Karney, 2012).

Marital Resources

Just as individuals have stable characteristics that are likely to influence their responses to deployment, relationships also have stable characteristics that are likely to have an impact on how the relationship is affected by deployment. The conceptual model treats these characteristics of the relationship between the service member and the spouse as a separate set of preexisting conditions that may explain variability in family outcomes across the deployment cycle. Specific constructs measured within this theme include marital history, ordering of marriage and military service, and parental status:

- *Marital history.* Aspects of marriage associated with lower divorce rates in the civilian population are likely to be associated with more-stable marriages across the deployment cycle. At baseline only, service members and spouses completed several items asking about their marital history, including the age of each partner at marriage, the length of the current marriage, the number of prior marriages, and (if applicable) how prior marriages ended. Prior research suggests that marriages of longer duration, between partners who married at older ages, and first marriages are more resilient than shorter marriages, marriages at younger ages, and remarriages (Karney and Bradbury, 1995).
- *Ordering of marriage and joining the military.* In couples married before one spouse joined the military, the demands of deployment may come as a surprise to the nonmilitary spouse. In couples that married after the service member joined, the demands of deployment may be more consistent with both partners' expectations of what the marriage would be like. To examine this potential modifier of the deployment's effects on family outcomes, the baseline assessment asked service members and spouses about the timing of the marriage relative to the service member joining the military.
- *Parental status.* Among civilian marriages, parental status affects marital outcomes in multiple ways. Sharing biological children is associated with lower divorce rates (White and Booth, 1985b), whereas the presence of children from prior relationships is associated with higher divorce rates (White and Booth, 1985a). The conceptual model suggests that parental status should similarly modify the effects of deployment, such that couples that share children together should weather the separation better than couples without children or couples with children from prior relationships. At baseline, spouses were asked to complete a household roster, indicating the presence, number, and ages of all children in the home. At subsequent assessments, spouses will be asked to indicate whether there are changes in the household and, if so, will complete the household roster again. At every assessment, service members and spouses are also asked whether the service member or spouse is pregnant or trying to get pregnant, whether the pregnancy was planned, and (if applicable) when the service member or spouse is due.

Nonmilitary Circumstances

The conceptual framework highlights the nonmilitary circumstances of each family as a third set of preexisting conditions that should modify deployment's effects on well-being. Although deployments are likely to be stressful for all who experience them, some families manage deployment-related stress in a context that is otherwise free from major stresses and rich in support and resources, whereas others must manage the demands of deployment in addition to other serious demands on their time. Families surrounded by greater sources of support and lower levels of stress are likely to be more resilient than families with fewer sources of support or with higher levels of stress. Specific constructs measured within this theme include family income, economic strain, acute (nonmilitary) stress, and social support:

- *Family income.* Money is a fungible resource that can be applied toward coping with almost any source of stress. The conceptual model suggests that families with higher levels of household income may have the resources to cope more effectively with the demands of deployment. Both service members and spouses are asked about their sources of individual and household income at every assessment. Note also that service members and spouses report changes in expenses during deployment; these items are discussed more under "Predeployment" later in this chapter.

- *Acute (nonmilitary) stress.* Among civilian families, the experience of acute stress has been associated with poorer individual well-being and less satisfying and stable relationships (Karney, Story, and Bradbury, 2005). Moreover, the negative implications of acute stress are magnified for couples that are simultaneously dealing with other enduring demands on their energy and attention (Rauer et al., 2008). Among military couples, the conceptual framework suggests that the implications of deployment may be similarly magnified within couples that are simultaneously coping with other stressful events in their lives. To assess these nonmilitary stressors, service members and spouses are asked, at every assessment, 15 items adapted from "The List of Threatening Experiences" (Brugha et al., 1985). For this measure, spouses indicate whether they have experienced each possible stressor (e.g., the death of a parent or child).
- *Social support.* Social support has consistently been associated with better physical, mental, and emotional health (Taylor, 2011) and has been found to mitigate the deleterious impact that stress can have on health and well-being (Cohen and Wills, 1985). This is true regardless of whether the support is instrumental (e.g., a loan, babysitting) or expressive (e.g., talking to someone about a problem), although expressive, or emotional, support is generally found to be more protective. As a result, the presence of social support should alleviate deployment-related stress. Three separate measures tap into different aspects of support, including sources of social support, the presence of instrumental support, and the presence of expressive support. All of these items are assessed at baseline and all subsequent survey waves:
  - *Sources of support.* Service members and spouses are asked how much they are able to draw on nine different sources of support (e.g., own family, spouse's family, other military families).
  - *Instrumental support.* Instrumental support is assessed through four items that ask the service member and spouse whether they could count on receiving help with housing, medical care, child care, and finances if those resources were needed. These items have been used in existing studies, including the Fragile Families and Child Wellbeing Study (Reichman et al., 2001), the Three-City Study (Angel et al., 2009), and FPNMAD (e.g., Trail and Karney, 2012).
  - *Expressive support.* Expressive support is assessed via one item that assesses whether service members, spouses, and study children have someone to listen to their problems if needed.

Prior Military Circumstances

A family's prior experiences in the military are another set of preexisting conditions that may shape their reactions to a new deployment. The conceptual framework allows for two types of effects. On the one hand, families that are accustomed to the demands of military life may have developed strategies to cope with those demands, leading to greater resilience. On the other hand, families that have experienced negative consequences of military stress in the past may have fewer resources to cope with a new deployment, leading to greater vulnerability.

The Deployment Life Study will be able to examine both of these possibilities. Specific constructs measured within this theme include general characteristics of service, deployment history, prior combat experiences, expectations for deployment, relocation, commitment to the military, and job satisfaction:

- *Characteristics of service.* Deployments are stressful for all service members, but the demands, expectations, and available resources also vary significantly across segments of the military. For example, deployment length varies across the services in terms of whether or not a service member is typically deployed with his or her unit (e.g., airmen are often deployed in individual augmentations, not with the unit with which they trained). To evaluate any differences in the impact of deployments across different parts of the military, each service member was asked at baseline to report his or her service (e.g., Army, Navy, Air Force, Marine Corps), component (e.g., active, reserve), length of service, and pay grade.
- *Deployment history.* As a consequence of the intense pace of deployments during the past decade, a high proportion of the current military has already experienced one or more deployments. Families that know what to expect may cope with a new deployment differently from families that have not been separated by deployments before. To evaluate these differences, service members and spouses who also indicate they are in the military are asked about the number, length, and timing of any prior deployments they had experienced. At follow-up assessments, service members and eligible spouses are asked only about the impending or current deployment.
- *Prior combat experiences.* All deployments are stressful, but all deployments are not stressful in the same way. One aspect of deployments that has been directly linked to service members' mental health outcomes in past research is whether or not the deployment exposed the service member to combat (Hoge, Auchterlonie, and Milliken, 2006; I. Jacobson et al., 2008). Service members who indicate that they have experienced at least one deployment of at least one month in duration outside the continental United States (OCONUS) are asked about their experiences with combat. We use the Hoge, Castro, et al. (2004) combat-experience scale, which has been frequently used in research among those serving in the military. It asks respondents to indicate whether or not they have ever experienced a range of combat-related scenarios (e.g., received small-arms fire, knew someone who was seriously injured or killed, was wounded). In follow-up waves, the survey asks similar questions about deployment experiences but focuses on the service member's most recent or current deployment.
- *Expectations for deployment.* Military families that have realistic expectations for the demands of deployment are in a better position to make realistic preparations for those demands and so may function more effectively across the deployment cycle. In contrast, families with unrealistic expectations (i.e., those with expectations that are too rosy or too pessimistic) may not prepare as effectively, either because they do not see a need for concern or because anxious anticipation leaves them overwhelmed. In the baseline survey, service members, spouses, and study children were asked whether the amount of time the service member has spent away from home in the past 12 months was more or less than what the respondent expected. In subsequent survey waves, service members who are not currently deployed are asked whether they know the dates and location of an upcoming deployment, how long he or she expects to be gone, whether the deployment will involve

engagement with a combatant, and how frequently the service member expects to communicate with his or her family. Spouses are asked similar questions, focusing on dates and location of the upcoming deployment and expectations about communication with their service member spouses.

- *Relocation.* As they anticipate an impending deployment, families that have recently been required to move may not have as rich or as strong a network of local support in place as families that have been living in the same place longer. At baseline, service members and spouses were asked about their prior experience of mandated relocations, known as a permanent change of station (PCS). At each subsequent assessment, both partners are asked to indicate whether they have experienced a move to a new location. All waves ask whether respondents live on or off base and, if off base, whether housing is rented or owned (e.g., they pay a mortgage) and how far from the base it is.
- *Commitment to the military.* Deployments require sacrifices from all members of a military family. Making those sacrifices is likely to be more meaningful, and thus easier, for service members and spouses who are more deeply committed to the goals and institution of the military. To assess commitment to the military, service members and spouses receive five items used in DMDC's Status of Forces Survey (SOFS) and from O'Reilly and Chatman's (1996) assessment of organizational commitment. At baseline and all follow-up assessments, service members and spouses are asked how much they agree with such statements as "You are willing to make sacrifices to help [branch of service]" and "I am glad I am/my spouse is part of my service." Study children also receive three items on commitment to the military. These items assess how much the child agrees with three statements: "Being a child in a military family inspires you to do the best job you can," "You are willing to make sacrifices so that your family can continue to contribute to the military," and "You are glad to be part of a military family." These items are asked at baseline and in each subsequent survey.
- *Job satisfaction.* Service members' and spouses' overall evaluations of their satisfaction with their jobs are preexisting conditions of the family, but job satisfaction is also an important outcome for the Deployment Life Study. Satisfaction is likely to vary as a function of personal experiences during deployment, and, in turn, it is likely to predict outcomes subsequent to deployment—most notably, decisions about remaining in the military (Farkas and Tetrick, 1989). Accordingly, this construct is measured with a single item at every assessment. Service members and spouses, because they have different roles, receive slightly different versions of this item. Each service member is asked how satisfied he or she is with the military way of life. Each spouse is asked how satisfied he or she is with the quality of his or her life as the spouse of a service member. Both items are also taken from DMDC's SOFS.

Adaptive Processes

The conceptual framework informing the Deployment Life Study suggests that the conditions of military families prior to a deployment can affect their outcomes during and postdeployment only through direct effects on the ways in which military families *adapt* to the deployment. Some preexisting conditions (e.g., a strong marriage, rich social support networks, absence of prior trauma) should facilitate effective adaptation, whereas other preexisting conditions (e.g., depressive symptoms, recent relocations, other sources of stress) should make effective adaptation more difficult. As noted in Chapter Two, *adaptation* is broadly defined to include all of the ways in which family members prepare for and manage the demands of deployment and running the household and raising children during the deployment. Adaptation may vary depending on the stage of deployment process—predeployment, during deployment, or postdeployment. Thus, in this section, we demarcate our discussion of adaptive processes by (roughly) when they occur in the deployment cycle. Many of the constructs and measures described in this section have different meanings at different stages.

Specific constructs measured within this theme include quality of communication between spouses, parent–child communication, intimate-partner violence, managing daily tasks, readiness (including combat, family, and child), socialization with other military families, and utilization of support programs.

Predeployment

The predeployment phase of the deployment cycle is the time in which service members and their families were informed of an impending deployment and are preparing for it. The characteristics and activities of families during this period can set the stage for how well, or how poorly, the family will adapt during a deployment:

- *Quality of communication between spouses.* A central feature of the way in which couples adapt to their circumstances is the way they communicate with each other. Marital communication features prominently in all models of marital success and failure (Bradbury and Karney, 2010) and is a consistent predictor of divorce and of change and stability in marital satisfaction (e.g., D. Johnson, 2005). The behaviors exchanged between spouses are difficult to assess via self-reports because spouses are highly inaccurate reporters on their own and each other's behavior (N. Jacobson and Moore, 1981). Instead of asking about their communication behaviors, the Deployment Life Study asks service members and spouses at every assessment to report on their affective responses to their interactions with each other. To measure these responses, eight items were adapted from the Positive and Negative Affect Schedule (PANAS) short form, a frequently used measure of positive and negative affect (Watson, Clark, and Tellegen, 1988; Thompson and Cavallaro, 2007). For these items, each respondent rates the degree to which he or she feels several positive and negative emotions after interacting with his or her spouse.
- *Parent–child communication.* Just as communication is central to spousal relationships, it is central to parent–child relationships as well. To assess this construct, the Deployment Life Study takes two approaches. First, spouses and children both receive versions of the PANAS scale at every assessment to measure parents' and children's affective reactions to communication with each other. Second, each child assessment also includes the ten-item Parent–Child Communication Scale (Loeber and Stouthamer-Loeber, 1998), which requires child respondents to answer questions about how they communicate with each of their parents (e.g., "Can you let your parent know what is bothering you?").
- *Intimate-partner violence.* Whereas good communication can be an effective adaptive process, physical aggression between partners is a maladaptive process, but one that may be exacerbated by stresses before, during, and after deployments. Intimate-partner violence is measured using three items each from the physical-assault and psychological-aggression subscales from the Conflict Tactics Scale 2 (CTS-2) (Straus et al., 1996). For each of these

items, respondents report the frequency with which they or their partners perform aggressive behaviors (e.g., insulting or swearing at partner, pushing or shoving partner). The measure is administered to service members and spouses at every assessment.

- *Managing daily tasks.* Because the deployment of one spouse requires the other spouse to take full responsibility for managing their home, the Deployment Life Study asks spouses of service members to complete, at every assessment, a ten-item measure of perceived ease or difficulty in managing daily tasks, adapted from a measure used in the Sample Survey of Military Personnel (SSMP).[2] These items ask respondents to rate how well they perform regular civilian responsibilities, such as working at their jobs and cleaning their homes.
- *Socialization with other military families.* As noted above, social support can moderate the negative effect of stress and strain. For military families, social support from other families that know firsthand what it is like to experience the absence of a loved one for an extended period of time, often to a dangerous place, may be particularly valuable in the stress-buffering process. The Deployment Life Study asks service members, spouses, and study children whether they or other members of their families routinely interact with other military families. This item is asked at baseline and each subsequent survey wave.
- *Use of military and nonmilitary support programs.* Similar to social support from military families, the use of military and nonmilitary support programs can act as a resource that families can use to cope with a deployment. A series of questions, asked of both service members and spouses, probes whether the family has used any such program, what program was used, and how recent the program utilization was. These items are asked at baseline and again at all subsequent interviews.
- *Readiness.* Readiness is a key construct among military families. Whether or not a service member, spouse, or children feels that he or she is prepared for a family member's deployment can be a protective factor helping him or her to adapt to additional stresses and strains associated with deployment. In contrast, families that do not feel prepared for such a tumultuous period in a family's history may be at a disadvantage when it comes to coping with a deployment. The Deployment Life Study asks about three types of readiness: combat, family, and child. In total, the survey includes 14 items from the Deployment Risk and Resilience Inventory (DRRI) (King et al., 2006). These items are also used in DMDC's SOFS:
  - *Combat readiness.* Combat readiness includes five items asking only service members how prepared they feel that they and their units are for an upcoming deployment. Specific items focus on manning, unit-level training, parts and equipment, physical fitness, and personal training and experience. These items were asked of all service members at baseline. In subsequent waves, they are asked only if a service member indicates that he or she is preparing for a deployment.
  - *Family and marital readiness.* Family readiness includes five items asked of both service members and spouses. The first three items ask whether the family will have money for food, rent, and other living expenses during the deployment; whether the family

[2] The Army Personnel Survey Office at the U.S. Army Research Institute for the Behavioral and Social Sciences conducts the SSMP semiannually in the spring and fall on behalf of the Deputy Chief of Staff, G-1. Each survey contains demographic items and specific items on key topics (e.g., well-being, quality of life, job satisfaction, career matters, career intent, morale, readiness, unit climate, and family matters). The survey is longitudinal (i.e., a repeated cross-section) so that trends in these key issues can be tracked over time. More information about the SSMP can be found at U.S. Army Family and Morale, Welfare and Recreation Programs (undated).

has developed a financial emergency plan; and whether the service member or spouse has life insurance. The next two items refer specifically to the couple and ask whether the service member or spouse has talked to his or her spouse about how the upcoming deployment may affect the marriage and whether either party has talked with a professional (e.g., counselor) about how the deployment may affect the marriage. These items are asked at baseline and again in subsequent waves if a service member or spouse indicates that the family is preparing for an upcoming deployment.

– *Child readiness.* In households with children, both service member and spouse are asked four items about how well the child or children in the household have been prepared for an upcoming deployment. Items ask about whether the service member or spouse has talked with a professional about how the deployment may affect the children, talked directly with the children about what to expect during the deployment, connected the children with a support group (e.g., Operation Purple Camp), and connected with other military children in the community. These items are asked at baseline and again in subsequent waves for families expecting a deployment.

During Deployment

The conceptual framework guiding the Deployment Life Study treats the characteristics of a given deployment as a mostly exogenous variable. Because service members cannot control when and where they are deployed, the nature of their deployments should be independent of the situation and characteristics of the service member and his or her family prior to a deployment. Nevertheless, the nature of a deployment could have direct and indirect effects on a family's immediate and long-term outcomes postdeployment. The direct effects include consequences that can be attributed to experiences during deployment (e.g., physical and emotional injuries). The indirect effects include consequences that are mediated by adaptive processes within the family (e.g., the stress of deployment can lead to emotional distance between spouses, which, in turn, can lead to marital distress or divorce). By measuring aspects of the family's adaptive processes, as well as concrete characteristics of the deployment itself, the Deployment Life Study will be able to tease apart the degree to which family members' own actions may buffer or exacerbate the effects that their deployment experiences can have on their outcomes.

At each wave for which a service member reports that he or she is currently deployed, the Deployment Life Study asks a series of questions about what life is like for both the service member and his or her family. For service members, these items include information about the length of the current deployment, communication with his or her family (e.g., frequency, means, topics discussed), difficulties with family, and general military-related stressors (e.g., experiencing physically demanding tasks, lack of sleep, working under pressure). The military-stressors measure includes six items derived from Zohar et al.'s (2004) Military Life Scale—in particular, the subscale on military tasks and demands. Other items were developed specifically for the Deployment Life Study.

Spouses are asked similar questions about the length of the deployment and communication. They are also asked a series of 11 items about problems the family may be experiencing during deployment (e.g., feeling like people in the community do not understand your problems, losing contact with friends or family, having more responsibilities in child care). Five additional items ask each spouse about how the deployment may have affected his or her life (e.g., "You like to keep track of the news about the war" and "You worry about your spouse while he or she is deployed"). Eight items ask spouses how true they believe statements describing their circumstances during the deployments are (e.g., "I feel proud," "I feel I have more responsibilities at home," "I spend significantly less time on work outside the home"). Nine items ask spouses how much they rely on different sources of support during the deployments (e.g., own family, service member's family, coworkers). Finally, 11 items ask the spouse how he or she believes that the focal (or study) child is reacting to the deployment (e.g., feels worried about his or her military parent, is mature, spends more time alone). Many of these items were developed specifically for the Deployment Life Study or adapted from prior RAND work (see Chandra, Burns, et al., 2008; Chandra, Lara-Cinisomo, Jaycox, Tanielian, Burns, et al., 2010).

Study children receive one item that asks how long they expect their military parents to be away from home, once those parents are notified of a deployment. They are also asked how often they expect to communicate with their parents, as well as whether or not their parents have discussed the deployment with them. Subsequently, once the service member is deployed, the study child is asked how often he or she communicates with his or her deployed parents, the means by which they communicate, and the topics of their conversations. Study children are asked, like service members' spouses are, about problems they or their families may be having during the deployment (e.g., "You rely on your friend more for help and support," "You get into more trouble at school") (Chandra, Burns, et al., 2008; Chandra, Lara-Cinisomo, Jaycox, Tanielian, Han, et al., 2011). They also receive a list of ten experiences they may have had during their parents' deployments (e.g., "missed school because you don't feel like leaving home," "have a hard time dealing with life without your [parent]") (Chandra, Burns, et al., 2008; Chandra, Lara-Cinisomo, Jaycox, Tanielian, Han, et al., 2011). Like spouses, study children are also asked six items about sources of support during a parental deployment. These items were developed specifically for the Deployment Life Study.

Reintegration

Reintegration refers to the period in which the deployed service member returns home and is reunited with his or her family. Perhaps surprisingly, little research has focused on this period of the deployment cycle. The Deployment Life Study is poised to make a major contribution in this regard by including measures that focus on this piece of the deployment cycle.

Both service members and spouses are asked how much time has passed since the service member returned from deployment. This will help us to assess whether the impact of deployment may dissipate over time. Each service member, spouse, and focal child is asked a series of items about how problematic the return of the service member has been with respect to several domains (e.g., getting to know the spouse or parent again, worrying about the next deployment, rebalancing responsibilities, worrying about parents getting along). Each spouse is asked a series of nine items about how the child has handled the return of the deployed parent. Each service member, spouse, and study child is asked a modified version of the Post-Deployment Reintegration Scale (PDRS) (Blais, Thompson, and McCreary, 2009) consisting of ten items about how much the respondent agrees with statements about the service member's deployment (e.g., "Since being deployed, my spouse/military parent has had difficulties understanding me," and "I feel closer to my spouse/military parent than ever"). Finally, both the spouse and study child are asked how well they think their spouse or military parent is adjusting to being home and how well they themselves are adjusting. Service members are also asked to assess how well the readjustment is going for them. With the exception of the PDRS, many of these items were utilized in early RAND work on children from military families (see Chandra, Burns, et al., 2008; Chandra, Lara-Cinisomo, Jaycox, Tanielian, Burns, et al., 2010).

Immediate Outcomes

Emergent Traits

The conceptual framework invokes the concept of emergent traits to acknowledge that the experience of a deployment can change people like few other experiences can. Service members, in particular, may be different people postdeployment from who they were predeployment, to the extent that they may be disabled, wounded, or traumatized. Other family members may also emerge changed, to the extent that deployments can be associated with changes in jobs, in education, in personality, and in other individual characteristics that might otherwise have been stable. These are immediate outcomes of deployment that may contribute to or detract from the long-term well-being of military families after deployment. Whether or not these changes occurred will be assessed simply by administering measures of any individual difference that can change both pre- and postdeployment. Specific constructs unique to this theme include life satisfaction, risky behavior, drugs, alcohol, tobacco, PTSD symptoms and TBI, and relationship quality:

- *Life satisfaction.* Life satisfaction can be thought of as a global measure of well-being (Diener et al., 1985; Pavot and Diener, 1993). Individuals who have experienced a deployment, either themselves or as a family member, and have emerged from the experience with relatively few negative consequences are expected to report greater life satisfaction than those who may not have been as resilient. Life satisfaction is measured with one item, asked at every survey, asking family members to indicate the level of satisfaction they have with their lives, taken as a whole, right now.
- *Risk behaviors.* Increase in risk behaviors (e.g., not wearing a seat belt, fighting, drinking and driving) may signal a maladaptive response to a deployment experience (Bray et al., 2010). Service members in the Deployment Life Study are asked about the frequency with which they engage in eight risk behaviors. These items are modified from a risk scale created for the National College Health Risk Behavior Survey (NCHRBS) (Brener, Collins, et al., 1995; see also Douglas et al., 1997, and Brener, Kann, et al., 2002). Questions about risk behavior are not asked of spouses but are asked of service members at all survey waves. Risk behavior (as well as drug, alcohol, and tobacco use) is also assessed among study children in the Deployment Life Study. A modified 15-item version of Farrell, Meyer, and White's (2001) behavior-problem scale is utilized at baseline and in all subsequent survey waves. The items focus on physical aggression (seven items), drug use (six items), nonphysical aggression (one item), and relational aggression (one item) by asking how often the study child has engaged in each behavior.
- *Drugs, alcohol, and tobacco.* Use of drugs, alcohol, or tobacco may also represent a maladaptive response to a deployment experience. Service members who may not have used or abused these substances before a deployment may change their habits upon return. The Deployment Life Study focuses on four types of substances: prescription drugs, alcohol, illegal drugs, and tobacco:

- *Prescription drugs.* The survey contains two items related to using prescription drugs that were not prescribed to the respondent or taken only for the feeling they caused. The first asks whether either had ever happened; the second asks whether either had occurred in the past 30 days. Both items are taken from the 2007 National Survey on Drug Use and Health (U.S. Department of Health and Human Services, 2008). Both service members and spouses receive these items at baseline and all subsequent survey waves.
- *Alcohol.* Three items assess the frequency and amount of alcohol consumption by service members and spouses. These items include the number of days the respondent has consumed alcohol in the past 30 days, the average number of drinks consumed on those days, and the number of days the respondent binge-drank (five or more drinks for men, four or more drinks for women) in the past 30 days. All items come from the 2007 National Survey on Drug Use and Health.
- *Illegal drugs.* Abuse of illegal drugs is assessed with a modified version of the CAGE questionnaire (Mayfield, McLeod, and Hall, 1974).[3] The measure typically assesses problem drinking. This version of the CAGE contains seven items that ask service members and spouses whether they have ever felt that they should cut down on drinking or drug use, whether others have criticized their drinking or drug use, whether they have ever felt guilty about their drinking or drug use, whether drinking or drug use has interfered with their daily lives, whether physical fights frequently result from drinking or drug use, whether drinking or drug use has frequently caused problems with friends or family, and whether the respondent has operated a car or other equipment while under the influence of drugs or alcohol. All items refer to the past four months and are asked at baseline and again at every survey wave.
- *Tobacco.* Tobacco use is assessed with three items that ask service members and spouses the number of days they smoked part or all of a cigarette (in the past 30 days), the average number of cigarettes smoked on those days, and the number of days they used chewing tobacco (in the past 30 days). All items come from the 2007 National Survey on Drug Use and Health.

• *Psychological health.* Recent reports suggest that 10–20 percent of veterans from OIF and OEF screen positive for PTSD (Ramchand et al., 2010). PTSD has been linked to negative outcomes for service members and their families, including decreased physical health, loss of work, and diminished well-being (Zatzick et al., 1997). Research and reports also indicate that approximately 19 percent of veterans report having experienced a probable TBI during deployment. Experiencing a TBI has also been linked to negative health outcomes (Hoge, McGurk, et al., 2008). In many ways, PTSD and TBI are the prototypical examples of enduring traits in that they can significantly affect what a service member is like pre- versus postdeployment and how his or her family will be affected by his or her return:
  - *PTSD.* The Deployment Life Study uses the PTSD Checklist (PCL) (Weathers, Huska, and Keane, 1991) to assess PTSD symptoms among service members. Respondents

[3] The abbreviation *CAGE* represents the four yes/no items that the original questionnaire asks: (1) Have you ever felt that you ought to *cut down* on your drinking? (2) have people *annoyed* you by criticizing your drinking? (3) have you ever felt bad or *guilty* about your drinking? and (4) have you ever had a drink first thing in the morning to steady your nerves or to get rid of a hangover (*eye-opener*)?

indicate how much they have been bothered by 17 different symptoms in the past 30 days. Only service members who indicate that they had, in fact, experienced a traumatic event receive the full PCL. The PCL has been used in prior work on military populations (see Hoge, Castro, et al., 2004; Hoge, Terhakopian, et al., 2007; Bliese et al., 2007; Tanielian and Jaycox, 2008); the version used in the Deployment Life Study is the specific PCL (PCL-S). PTSD symptoms among service members are assessed at baseline and in all subsequent surveys. Probable diagnoses of PTSD were derived using guidelines offered by Weathers, Litz, et al. (1993). Respondents are considered to have probable PTSD if they report being at least moderately bothered in the past 30 days on at least one of the reexperiencing items, three of the avoidance items, and two of the arousal items. This scoring has been shown to have high specificity and sensitivity (see Brewin, 2005, for a review of different scoring methods).

  - The survey also assesses PTSD symptoms among spouses using the Primary Care PTSD Screen (PC-PTSD) (Prins et al., 2004). This measure uses a screen item to "cue" respondents to traumatic events and then asks whether four specific symptoms have occurred in the past 30 days. It is important to keep in mind that this measure is not directly comparable to the Primary Care Checklist–Military Version (PCL-M). PTSD symptoms among spouses are assessed at baseline and all subsequent surveys. Spouses are considered to screen positive for PTSD if they endorse two or more of the four symptoms.
  - *Probable TBI. Probable TBI* is defined as having experienced an injury that resulted in losing consciousness, memory loss, being dazed or confused, or "seeing stars" (Schwab et al., 2007). Four items adapted from the Brief Traumatic Brain Injury Screen (BTBIS) (Schwab et al., 2007) are used to assess whether or not service members (not spouses) have possibly experienced TBI. The first two items assess whether or not a TBI-related injury has ever occurred (e.g., losing consciousness) and how recently the injury occurred. The third asks whether certain symptoms began or increased in severity after the injury (e.g., ringing in the ears, sensitivity to light). The fourth asks whether any of those symptoms has occurred in the past week. These items are asked at baseline, and subsequent survey waves assess probable TBI-related injuries experienced in the past four months.

- *Relationship quality.* The quality of the relationships among family members is one of the central outcomes of interest in the Deployment Life Study, but it is also a preexisting condition of the family. Some families approach each deployment with strong ties among family members, whereas others have weaker ties or relationship distress. The longitudinal design of the Deployment Life Study will allow for analyses of change in the quality of family relationships (as an outcome), controlling for the quality of those relationships prior to the deployment (a preexisting condition). Specific constructs measured within this theme include marital satisfaction, parenting satisfaction, and quality of family environment:
  - *Marital satisfaction.* Marital satisfaction is one of the most frequently studied outcomes in all research on marriage (Karney and Bradbury, 1995). Couples that are more satisfied with their marriages are healthier, less likely to divorce, and live longer than couples that are unsatisfied with their marriages (Kiecolt-Glaser and Newton, 2001). In the Deployment Life Study, marital satisfaction is being measured with nine items adapted from a set that has been used previously in the Florida Family Formation

Survey (Rauer et al., 2008). It assesses relationship satisfaction in nine areas, including satisfaction with time spent together, communication, and trust, but all of the items cohere as a reliable, single-factor scale. This measure is administered to spouses and service members at every assessment.

- *Parenting satisfaction.* For military couples who are also parents, the Deployment Life Study is also measuring satisfaction with parenting, a construct that has also been associated with psychological well-being and health outcomes in parents, as well as their children (Anthony et al., 2005; Crnic and Greenberg, 1990). To measure this construct, parents at all assessments receive a six-item measure asking them to rate their level of agreement with statements about parenting (e.g., "Being a parent is harder than I thought it would be"). This measure is adapted from a nine-item measure of parental aggravation developed for the Child Development Supplement of the Panel Study of Income Dynamics (PSID) (Abidin, 1995; see also Hofferth et al., 1997, and PSID, undated).
- *Quality of family environment.* As an additional measure of parents' satisfaction with family relationships, service members and spouses who are parents, as well as their study children, receive at every assessment a six-item measure of family environment that assesses how the family functions as a unit. These items were selected from the cohesion and conflict subscales of the Family Environment Scale (FES) (Moos and Moos, 1994) and ask respondents to rate how well statements (e.g., "We fight a lot in our family") describe their families.

Long-Term Outcomes

Although the immediate outcomes described above are important to assessing overall family functioning, especially as it may be affected by deployment, the ultimate goal of the Deployment Life Study is to assess the *long-term* experiences of families that go through a deployment. Immediate outcomes can set the stage for longer-term consequences, either as manifestations of healthy adaptations to deployment (e.g., high-quality family interactions, positive life satisfaction, responsible use of substances) or as evidence of maladaptive responses (e.g., marital discord, substance abuse). Thus, the long-term outcomes of the families in the Deployment Life Study reflect not only immediate outcomes but also all of the characteristics, attributes, and experiences described in this chapter and depicted in the conceptual model shown in Chapter Three. In this section, we identify six areas of long-term well-being measured in the Deployment Life Study: *marital dissolution, military retention, child's academic achievement, financial well-being, physical health,* and *mental, emotional, and behavioral health.*

Marital Dissolution

Perhaps the ultimate long-term negative outcome for a military family is marital dissolution. There are two ways to measure marital dissolution. The first is simply to ask whether or not a divorce (or separation) has occurred. At all waves, both service members and spouses are asked whether they are still married to each other. The second way to measure marital dissolution is to assess the self-reported likelihood that dissolution will occur. The Deployment Life Study uses a three-item variant of the Marital Status Inventory (MSI) to assess how likely it is that a couple will separate (Weiss and Cerreto, 1980). These items ask each service member and his or her spouse whether he or she has independently thought about divorce or separation (e.g., who would take custody of the children), whether he or she had made plans to discuss divorce or separation with his or her partner, and whether he or she had contacted a lawyer to make preliminary plans to divorce.

Retention Attitudes

From the perspective of DoD, retention is a key long-term outcome. The all-volunteer force is dependent on service members choosing to remain in their military careers beyond their required terms of service. A family's experience with deployment very likely affects this decision. The Deployment Life Study contains several items about retention intentions, among service members (both active and reserve components) and family members. For active duty, three items ask whether the service member would stay on active duty if given the choice, would stay until retirement-eligible at 20 years of service, and what his or her expected total years of service will be. Among reserve service members, one item assesses career intentions (e.g., leave after completing obligation, stay until mandatory retirement). A final item asks the service member whether his or her spouse favors staying in the military. Among spouses of both active and reserve component service members, one item asks how long the respondent expects the service member to serve. A second item asks how much he or she favors the service member staying in the service. All of the retention items are asked at baseline and again at all survey waves. All items for service members and spouses come from DMDC's SOFS. A single item, created specifically for the Deployment Life Study, asks the study child whether he or she plans to have a career in the military. Study children also receive an item asking them whether they favor their military parents staying in the military. This item is identical to the item that the service member's spouse receives.

Child Academic Achievement

A large body of research suggests that children may be negatively affected by the deployment of a parent (e.g., Chandra, Martin, et al., 2010). The Deployment Life Study assesses the impact that deployment can have on children across several important domains, including academic achievement. The survey focuses on educational aspirations, falling behind in school, problems in school, and academic engagement:

- *Educational aspirations.* Both the spouse and the study child are asked how far the child expects to go in school. This item, also used in the National Household Education Survey (NHES) (National Center for Education Statistics, undated), is asked at baseline and in all follow-up surveys.
- *Falling behind in school.* Three measures assess a child's ability to keep up with his or her peers academically. The first is grades (e.g., mostly As or Bs), which is assessed by both the spouse and the study child at baseline only. The second is a single item that asks both spouse and study child to indicate whether the child's schoolwork is excellent, above average, average, below average, or failing. It is asked at baseline and all subsequent waves. The third is asked only of spouses who indicate whether the study child has ever repeated a grade and what grade it was. This item was asked only at baseline. All items are used in the NHES.
- *Problems in school.* Three items, asked only of the spouse, assess the extent to which a child's performance or behavior at school has been brought to the attention of his or her

parents. Three items, from the NHES, ask spouses whether the child's school has called about behavior problems or problems with schoolwork and whether the child has shown significant improvement in some activity at school. These items are asked at baseline and all subsequent survey waves. Four additional items, asked only of the study child, ask how often the child has gotten in trouble at school, how often he or she has been sent to the office, how many times he or she has been suspended, and how often his or her parents have been contacted by the school because he or she has gotten into trouble. These items are asked at baseline and all subsequent survey waves. These items were created for the Deployment Life Study.

- *Academic engagement.* Academic engagement predicts a range of other academic outcomes, such as grade-point average (GPA), standardized test scores, and school dropout (Fredricks, Blumenfeld, and Paris, 2004). A seven-item engagement scale developed by Rosenthal and Feldman (1991) asks children how frequently they do their homework, prepare for class, and engage in other similar activities. These items are asked only of the study child, at baseline and all subsequent survey waves.

Financial Health

Depending on their lifestyles, expenses, and expectations, military families may perceive varying levels of economic strain, independent of their actual household income. Poorer coping across a deployment cycle may also be associated with greater levels of financial strain. To assess this construct, service members and spouses completed four items adapted from Gutman and Eccles (1999) at baseline and again at every subsequent assessment. These items assess whether the family could make ends meet, whether there has been enough money to pay bills, whether there was any money left over at the end of the month, and the level of worry caused by the family's financial situation.

Physical Health

Physical health is an important long-term outcome among military families, especially service members who may experience above-normal physical wear and tear as a result of both day-to-day military duties and deployments (Kline et al., 2010; Songer and LaPorte, 2000). Four items assess physical health, as used in the National Health Interview Survey (NHIS) (see Idler and Angel, 1990; Centers for Disease Control and Prevention, 2013), that were derived from the 12-item Short Form Health Survey (SF-12) (Ware, Kosinski, and Keller, 1995). These items focus on the degree to which the respondent—service member, spouse, or study child—is limited by his or her physical health, whether his or her health status limits his or her ability to work or attend school, how much physical health interferes with normal social activities, and the respondent's general physical health. This last item is typically referred to as *self-rated health*. Self-rated health is an excellent measure of global health and well-being and is a predictor of illness and mortality (Lorig et al., 1996).

Mental, Emotional, and Behavioral Health

Perhaps the largest set of long-term outcomes in our study consists of variables related to mental, emotional, and behavioral health. There is good reason to suspect that deployments have an impact on these types of outcomes because prior research has shown that rates of depression and generalized anxiety are higher among combat veterans than among those who have not experienced combat (Hoge, Castro, et al., 2004). Existing research among military families has shown that children whose parents deploy also report higher levels of depression and anxiety (Chandra, Martin, et al., 2010). Because the Deployment Life Study follows military families over the course of the deployment cycle, it will be able to assess changes across a variety of mental, emotional, and behavioral health constructs over time as service members prepare for deployment, deploy, and then return. Those constructs include overall mental health and well-being inventories, depression, anxiety, and mental health service utilization:

- *Overall mental health and well-being.* The Strengths and Difficulties Questionnaire (SDQ) is a 25-item assessment of child behavior in five areas: emotional problems (e.g., "You have many fears"), conduct problems (e.g., "You fight a lot"), hyperactivity (e.g., "You get easily distracted"), peer relationships (e.g., "You have at least one good friend"), and prosocial behavior (e.g., "You try to be nice to other people") (Goodman, 1997, 2001). The SDQ has been used to screen for child psychiatric disorders (Goodman et al., 2000). Both spouses and study children complete the assessment at baseline and all subsequent waves.
- *Anxiety.* Anxiety symptoms among study children are assessed using the five-item Screen for Child Anxiety Related Emotional Disorders (SCARED) (Birmaher, Khetarpal, et al., 1997; Birmaher, Brent, et al., 1999). The items include a single measure for five underlying factors: panic, generalized anxiety, separation anxiety, social phobia, and school phobia. The study child is asked the SCARED items at baseline and at each subsequent follow-up. Spouses are also asked the SCARED items at baseline and follow-up, either about the study child or, if there is no study child, the focal child. Anxiety among adults is measured using the anxiety subscale of the Mental Health Inventory 18 (MHI-18) (Sherbourne et al., 1992). This subscale is generally seen as an assessment of general anxiety. Both service members and spouses receive the four items in the anxiety subscale at baseline and at all subsequent survey waves.
- *Depression.* The eight-item Patient Health Questionnaire (PHQ-8) is a brief assessment of the severity of depressive symptoms, based on the *Diagnostic and Statistical Manual of Mental Disorders*, fourth edition (DSM-IV) criteria for depressive disorders (American Psychiatric Association, 1994; Kroenke, Spitzer, and Williams, 2001; Kroenke, Strine, et al., 2009). The items ask about symptoms, such as hopelessness, energy levels, appetite, concentration level, and lethargy, and the frequency with which they are experienced. Both service members and spouses receive the PHQ-8 at baseline and at all follow-up survey waves. *Probable depression* is defined by scores of greater than or equal to ten on the summed PHQ-8 index (Kroenke, Strine, et al., 2009). The adolescent version of the PHQ-8, the PHQ-A, is a 12-item measure used to assess depression among children and adolescents (J. Johnson et al., 2002). Provisional diagnoses of major depressive disorder (MDD), dysthymic disorder, and minor depressive disorder can be made from the PHQ-A. Similar to the scale for adults, it includes items about feeling sad and upset, having little interest in doing things, and trouble with sleep, as well as items about the frequency with which symptoms are experienced. Study children receive these items at baseline and all subsequent waves. Spouses are also asked the PHQ-A for the focal child.
- *Mental health service utilization.* Service members and spouses receive five items about their use of mental health services. These items are intended to assess whether services are utilized (including medication), whether a mental health service was needed (as self-reported by the respondent), what type of service was utilized (e.g., psychologist, chaplain), and why services were not used (among those who indicated a need). These items

will allow the study team to assess gaps in service utilization. All five items are included in the baseline survey, as well as all subsequent surveys. Study children receive four similar items but do not answer the item about why services were not received. Spouses receive the same four items with respect to the study child (or, if no study child is available, the focal child). All child mental health service utilization items are asked at baseline and again at all subsequent follow-up waves.

Summary

This chapter presented the main constructs and measures of the Deployment Life Study, informed by the conceptual model described in Chapter Three. Again, many of the measures described here could be considered part of more than one construct, depending on the research question. This flexibility will allow the data to be used across a wide array of research questions, both in the cross-section (i.e., using only the baseline data) and longitudinally. The scope of topics covered in the survey also increases the flexibility of the data and will allow us to address research question that, before now, could not be addressed using any existing survey data. The next chapter provides an overview of the baseline sample of the Deployment Life Study, focusing on many of the key constructs and measures outlined in this chapter.

CHAPTER SIX

The Baseline Sample

This chapter provides a description of the baseline Deployment Life Study sample. Screening and completion of the baseline survey for Army, Air Force, and Marine Corps families began in March 2011 and concluded in August 2012; completion of the baseline survey for Navy families occurred between November and April 2013. The baseline sample consisted of 2,236 households in which a service member and a spouse completed the baseline survey instrument and an additional 488 households in which a service member, spouse, and study child completed the survey. Thus, the total number of households in the baseline sample is 2,724: 2,724 service members, 2,724 spouses, and 488 children.[1] The chapter begins with basic sociodemographic characteristics (e.g., age, gender, education), then it moves to a description of the service member's military experience (and, in the case of dual-military couples, the spouse's military experience), and it concludes with a brief description of families' nonmilitary experiences (e.g., spousal employment, financial stress). Unless otherwise indicated, all data presented in this chapter use the sample weights described in Chapter Four.

Before we begin, four caveats are worth noting. First, the reader should keep in mind that the data presented in this chapter are meant to provide only a broad picture of the baseline sample. The data are not intended to present a picture of family readiness in the baseline sample. Second, and related, although we recognize that differences may exist across both service (Army, Air Force, Navy, and Marine Corps) and component (active and reserve) we do not disaggregate at those levels here. Future work will explicitly acknowledge and model these differences. Third, the reader should keep in mind that the ways in which the data are presented (e.g., as a dichotomous variable or as a sum score) are not necessarily how each construct will be measured in future analyses. However, when a standard scoring system is available, such as for probable PTSD, probable depression, and probable TBI, those operationalizations are used. And fourth, although comparing the results from the baseline Deployment Life Study sample and those of some comparable cohort would be desirable, it is not clear what that cohort should be. For example, there are many potential groups of interest: military families that have not experienced deployment; military families that experienced a deployment in prior conflicts; military families that experienced a deployment outside of OIF, OEF, or OND; and civilian families. Indeed, the appropriate comparison group will likely vary depending on the specific research question being examined. Comparing data from this study with data from other

[1] These 488 children are referred to as *study children* because the child him- or herself completes a unique survey. It is important to keep in mind that the 2,236 households made up of surveys from only a service member and a spouse may indeed have a child in the household. However, in these households, a child did not complete his or her own survey or was unable or ineligible to participate. In these households, the child is referred to as a *focal child*. Study children are always focal children; however, focal children are not always study children.

samples may also be limited because, as noted in Chapter Five, not all the measures used in the baseline survey are available in other studies conducted on those potential comparison samples. Thus, in this chapter, any generalizations made about the baseline sample are just that—general observations that do not make use of a comparison group.

Sociodemographic Characteristics

Overall, the service members in the sample are mostly male, white U.S. citizens in their early thirties. As shown in Table 6.1, 91 percent of service members in the Deployment Life Study baseline sample are male, and 76 percent of service members are white. The next-largest racial or ethnic group is Hispanic. The highest educational degree, attained by 54 percent of service members, is a high school degree or a high school degree with some college experience. Approximately 16 percent have associate's degrees or vocational diplomas; an additional 30 percent of

Table 6.1
Characteristics of the Baseline Deployment Life Study Sample: Service Members and Spouses

Characteristic	Service Member, n = 2,724	Spouse, n = 2,724	Study Child,[a] n = 488
Age, in years	32.8 (0.2)	32.1 (0.2)	13.6 (0.1)
Race or ethnicity (%)			
White, non-Hispanic	75.9	74.0	NA
Black, non-Hispanic	7.9	7.8	NA
Hispanic	12.2	11.9	NA
Asian	1.6	3.2	NA
Other	2.3	3.1	NA
U.S. citizen (%)	95.1	90.5	NA
Gender (% female)	8.6	91.4	50.3
Education (%)			
Less than high school	0.8	1.8	NA
High school diploma or equivalent	25.0	17.5	NA
Some college	29.0	27.9	NA
Associate's degree or vocational or technical diploma	15.8	18.2	NA
Bachelor's degree	19.6	26.6	NA
Master's, doctoral, or professional degree	9.8	7.9	NA
Enrolled in school (%)	26.8	24.7	NA
Current marriage			
Age at marriage, in years	24.9 (0.1)	24.2 (0.1)	NA
Length of marriage, in years	7.9 (0.2)	7.9 (0.2)	NA
First marriage (%)	81.5	82.3	NA

Table 6.1—Continued

Characteristic	Service Member, n = 2,724	Spouse, n = 2,724	Study Child,[a] n = 488
Children (%)			
Number of children			
None	NA	35.8	NA
One	NA	27.1	NA
Two	NA	24.9	NA
Three	NA	8.6	NA
Four or more	NA	3.5	NA
Mean, for those who have children	NA	1.8 (0.03)	NA
Age of children, in years, unweighted			
0 to 6	NA	52.9	NA
7 to 10	NA	23.1	NA
11 to 17	NA	23.2	NA
18 and above	NA	0.8	NA
Study children			
Enrolled in school (%)	NA	NA	85.7
Current grade (%)			
4–6	NA	NA	23.6
7–9	NA	NA	51.0
10–12	NA	NA	25.3
Household			
Mean number of adults	NA	2.2 (0.01)	NA

NOTE: NA = not asked. Values are weighted. Standard deviations are shown in parentheses where applicable.

[a] *Study child* refers to the child in the household between the ages of 11 and 17 who completed the child baseline survey.

service members have bachelor's degrees or higher; and almost 27 percent of service members were enrolled in school at the time of the baseline survey.

The demographic characteristics of spouses in the baseline sample mirror those of service members (see Table 6.1). Spouses are mostly female, white, and in their early thirties. Compared with service members, fewer spouses are U.S. citizens (91 percent versus 95 percent). A smaller proportion have high school degrees or some college experience (45 percent), and this difference is largely due to the fact that fewer spouses have high school diplomas or equivalent. About 25 percent of spouses were enrolled in school at the time of the baseline survey.

The family sociodemographic characteristics of the sample are also shown in Table 6.1. The majority of service members and spouses are in their first marriages, and their average age at marriage was around 25 years. They have been married an average of eight years, and about 64 percent of couples have one or more children, with an average of two children per couple.

About 53 percent of all focal children in sampled families were six years old or younger, and 46 percent were school-age (i.e., between seven and 17 years old) at the time of the baseline survey.

Finally, for study children between ages 11 and 17 who independently completed their surveys, the average age is around 14 years old. Most study children indicated that they were enrolled in school, with around 24 percent in grades four through six, 51 percent in grades seven through nine, and 25 percent in grades ten through 12. Thus, most study children were pre- or early adolescents at the time of the baseline survey.

Nonmilitary Experiences

As noted in Chapter Five, such factors as income and employment levels, financial stress, and stressful life events are risk factors for marriages. Experiencing financial strain or stressful life events may also exacerbate the impact that deployment stress can have on the family. Table 6.2 contains data describing the nonmilitary experiences of service members and spouses. All active component service members are considered to be full-time employed. However, reserve component service members may or may not have civilian jobs. Among reserve service members in our baseline sample (n = 721), more than half (58 percent) reported that they had full-time civilian jobs, with an additional 3 percent reporting having part-time civilian jobs. Among spouses, 39 percent reported that they were employed either part time or full time outside the

Table 6.2
Nonmilitary Experiences of the Baseline Deployment Life Study Sample

Characteristic	Service Member	Spouse
Employment status (%)[a]		
Full-time employed, 35+ hours per week	58.4	27.4
Part-time employed, <35 hours per week	3.0	11.9
Full-time homemaker	0.3	37.9
Student, whether full or part time	4.5	8.6
Unemployed, looking for work	6.6	7.4
Unemployed, not looking for work	12.6	2.9
Retired	0.2	0.3
Other, includes disabled	14.4	3.7
Personal annual income from nonmilitary earnings (%)[b]		
Less than $25,000	41.1	60.8
$25,000 to $49,999	24.6	24.1
$50,000 to $74,999	17.2	10.2
$75,000 to $99,999	7.6	2.4
$100,000 or more	9.5	2.5

Table 6.2—Continued

Characteristic	Service Member	Spouse
Household annual income from other sources (%)[c]		
Less than $25,000	NA	87.3
$25,000 to $49,999	NA	8.2
$50,000 to $74,999	NA	2.5
$75,000 to $99,999	NA	1.0
$100,000 or more	NA	0.9
Financial circumstances (%)		
Current financial condition		
Very comfortable and secure	33.4	33.8
Able to make ends meet without much difficulty	39.8	37.3
Occasionally have some difficulty making ends meet	16.8	18.5
Tough to make ends meet but keeping our heads above water	9.4	9.4
In over our heads	0.6	1.1
Financial stress scale[d]	1.9 (0.02)	1.9 (0.02)
Stressful life events, summed number of events	0.7 (0.03)	0.8 (0.04)

NOTE: NA = not asked. Values are weighted. Standard deviations are shown in parentheses where applicable. n = 2,724.

[a] Service member numbers refer to reserve component only (n = 721).

[b] Respondents selected their income levels from 21 categories ranging from "under $5,000" (1) to "greater than $100,000" (21). Here, these categories are truncated. Service member numbers refer to the reserve component only.

[c] Other sources include public assistance, child support, and family and friends.

[d] Mean of three items. Range is 1 to 4, with higher scores indicating more financial stress.

home. Almost 38 percent reported being full-time homemakers, 9 percent reported being full- or part-time students, and 7 percent were unemployed but looking for work.

Regarding family income outside of military pay, the focus is again on reserve component service members and all spouses. Among reserve component service members who reported pay outside the military, 66 percent reported earnings below $50,000 per year, just under one-fifth (17 percent) make between $50,000 and $74,999 per year, and the same proportion (17 percent) make more than $75,000 per year. Among spouses who reported receiving earnings from sources outside the military (n = 1,097), 61 percent earn less than $25,000. Just under one-fifth (24 percent) earn between $25,000 and $49,999, with the remaining 15 percent earning more than $50,000. In addition, about 13 percent of spouses reported receiving household income from other sources (e.g., child support, public assistance, from family and friends). The majority (88 percent) of this additional money is less than $25,000 per year. Service members are not asked this item.

Respondents also report their levels of financial stress and the number of stressful life events they experienced in the past year. One item asks both service members and spouses to assess their current financial situations. Roughly 33 percent of service members and spouses reported that they were very comfortable and secure in their financial situations. An additional 40 percent of service members and 37 percent of spouses reported that they were able to make ends meet without much difficulty. Approximately 20 percent of both service members and spouses indicated that they did sometimes have difficulties making ends meet, 10 percent said that it was difficult to make ends meet but they were managing, and about 1 percent said that they were in over their heads. Three additional items form a financial stress scale. Scores on this score range from one to four, with higher scores indicating higher financial stress. Mean scores are 1.9 for both service members and spouses.

Finally, when asked to report the occurrence-specific stressful events in the past four months (e.g., a serious illness, problems with police, a death in the family), on average, service members reported that almost two of these events had occurred. Spouses reported that just under one of these events had occurred.

Military Experiences

Service members and spouses both reported on their experiences in the military. Because 179 spouses indicated that they also served in the military, we include their experiences in a separate column in Table 6.3.

Table 6.3
Military Experiences of the Baseline Deployment Life Study Sample

Experience	Service Member, n = 2,724	Military Spouse,[a] n = 179
Service and component (%)		
Air Force		
Active	23.7	31.2
Guard	4.6	3.6
Reserve	2.8	4.9
Army		
Active	22.7	22.6
Guard	4.5	2.9
Reserve	1.6	4.4
Navy		
Active	24.7	20.6
Reserve	1.4	3.1
Marine Corps		
Active	13.4	6.8
Reserve	0.5	0.0

Table 6.3—Continued

Experience	Service Member, n = 2,724	Military Spouse,[a] n = 179
Pay grade (%)		
Officer		
O-1–O-3	10.5	14.2
O-4–O-5	7.2	4.8
O-6 and above	1.6	0.6
Warrant officer	1.9	1.9
Enlisted		
E-1–E-3	4.8	6.9
E-4–E-6	57.6	60.8
E-7–E-9	16.5	10.8
Years of active-duty service	9.8 (0.2)	NA
Live on base (%)	18.5	NA
Distance to base, in miles	32.5 (3.8)	NA
Relocation[b]		
Moves since 2004	2.9 (0.0)	NA
Months since last move	27.1 (0.6)	NA
Deployments		
Number of OCONUS deployments	3.2 (0.1)	2.4 (0.3)
Number lasting 1+ months[c]	9.6 (0.3)	7.8 (0.9)
Number since married	6.9 (0.2)	4.4 (0.6)
Aware of upcoming deployment (%)	54.2	NA

NOTE: NA = not asked. Values are weighted. Standard deviations are shown in parentheses where applicable.

[a] This column includes only those spouses who reported being in the military.

[b] 50 miles or more.

[c] *Deployment* here refers to any military service that required the service member to be away from home for one month or more at a time. It could include training.

Overall, across all branches, the majority of both service members and their spouses were serving on active duty.[2] A little more than 31 percent of the service members and 40 percent of the military spouses were serving in the Air Force; 29 percent of service members and 30 percent of military spouses served in the Army; 26 percent of service members and 24 percent of military spouses served in the Navy; and 14 percent of service members and 7 percent of military spouses served in the Marine Corps. The majority of service members and military spouses (79 percent) were enlisted.

[2] This is also true within branches of service.

The service members in the baseline sample were experienced soldiers, airmen, sailors, and marines. On average, service members had served in the military for almost ten years and had experienced almost three relocations (or PCSs) of 50 miles or more since 2004. On average, service members had experienced a little more than three OCONUS deployments, while military spouses had experienced a little more than two.[3] Overall, service members had experienced almost ten deployments lasting one month or more (including OCONUS and continental U.S. [CONUS] deployments),[4] and an average of seven deployments had occurred since the respondent had married the spouse in the survey. Military spouses experienced slightly fewer deployments lasting more than one month and fewer deployments occurring since marriage to their current partners. Furthermore, a little more than half of the service members reported that they had been notified of an upcoming deployment.[5]

Adaptive Processes

Table 6.4 shows descriptive statistics for baseline measures of family functioning and social support. Both service members and spouses reported high mean scores on the family environment scale and the parenting satisfaction scale, indicating that they believed that their families were well-functioning and that they were satisfied with their roles as parents. In addition, more

Table 6.4
Family Functioning and Social Support of the Baseline Deployment Life Study Sample

Characteristic	Service Member	Spouse
Family functioning		
FES[a]	2.8 (0.0)	2.8 (0.0)
Parenting satisfaction scale[b]	3.3 (0.0)	3.3 (0.0)
Social support (%)		
Perceived support[c]		
Place to live	85.3	84.2
Care if sick	83.4	81.2
Child care	75.9	74.5
Emergency loan	75.8	74.3
Someone to talk to	88.1	89.0

[3] These figures differ from those presented in Table 4.6 because the source of the data is different. Table 4.6 uses data provided by the services in the sample files. Table 6.3 uses self-report data from service members themselves.

[4] This particular item in the survey asks, "Since you joined the military, how many times have you been away from home for a month at a time or longer for any reason related to your military service?" Thus, the term *deployment* is broadly defined and could include something like training away from the service member's home base.

[5] The eventual dropping of deployment expectation as a necessary condition for recruitment in the screener may explain the low percentage of service members who reported that they had been notified of an upcoming deployment. It is also possible that the service member him- or herself was truly unaware of an upcoming deployment. Similarly, given the lag between receipt of the sample data from the services and actual implementation of the survey, it is possible that some service members had already deployed and returned. This question was not asked of spouses in the military.

Table 6.4—Continued

Characteristic	Service Member	Spouse
Sources of support (%)[d]		
Own family	68.5	66.6
Spouse's family	51.8	43.3
Civilian friends	30.4	45.3
Civilian employer	17.5	33.6
Fellow service members (or coworkers for spouse)	50.5	33.1
Other military families	34.3	39.8
FRG leader or unit POC	23.4	21.7
Nonmilitary community programs	12.3	9.0
Socialize with other military families	40.1	70.5
Use of resources (%)		
Uses medication for mental health problem	4.2	13.8
Needed counseling in past 4 months	13.6	16.0
Saw mental health specialist in past 4 months	73.1	68.1
Aware of military-sponsored programs	97.8	91.0
Family ever used program	58.3	54.9
Last time used program		
Within past week	10.5	13.2
Within past month	14.0	13.8
Within past 4 months	24.7	22.4
Within past year	50.7	50.6

NOTE: FRG = Family Readiness Group. Values are weighted. Standard deviations are shown in parentheses where applicable. n = 2,724.

[a] Mean of six items. Range of 1 to 3, with higher scores indicating better family functioning.

[b] Mean of six items. Range of 1 to 4, with higher scores indicating more satisfaction with parenting.

[c] The items ask the respondent to indicate whether he or she had each need, whether there are "enough people to count on" (1), "not enough people" (2), or "no one you can count on" (3). The table reports those who say there are enough people to count on.

[d] The items asked the respondent how much he or she drew on different sources of support: "very much" (1), "somewhat" (2), or "not at all" (3). The table reports those who reported drawing on each source "very much."

than 80 percent of service members and spouses reported that they had at least one person on whom they could count for support in finding a place to live, getting care if they were sick, and having someone to talk to (i.e., instrumental and expressive support). Almost two-thirds of service members and spouses indicated that they could find someone to help with child care or to give them an emergency loan.

Most service members and spouses indicated that they were "very much" able to rely on their family for support when needed. Just over half of the service members indicated that they could rely on their spouse's family for support, and 43 percent of spouses said that they could rely on the service member's family for support. About 51 percent of service members indicated that they could rely on their fellow service members for support, while 33 percent of spouses said that they could rely on their coworkers for support. About 45 percent of spouses said that they could rely on their civilian friends for support, compared with 30 percent of service members. Almost 40 percent of spouses and 34 percent of service members indicated that they could rely on other military families for support. Interestingly, 71 percent of spouses and 40 percent of service members indicated that they or other members of their immediate families socialized or communicated regularly with other military families. Roughly one-fifth of both service members (23 percent) and spouses (22 percent) indicated that they could rely on the service member's FRG leader or unit POC for support. Thus, the majority of service members and spouses said that they had social support networks, particularly within their families, although their sources of support outside of the family differed.

Finally, about 14 percent of service members and 16 percent of spouses reported needing counseling for emotional problems from a mental health specialist in the recent past, and almost 4 percent of service members and 14 percent of spouses reported that they had used medication for a mental health problem within the past four months. Of those who indicated a need for counseling, 73 percent of service members and 68 percent of spouses reported that they actually received counseling, suggesting that there are unmet needs for counseling among some military couples. The vast majority of service members and spouses indicated that they were aware of military-sponsored family support programs. Of those aware of the programs, most indicated that they had used the programs, with about half of those indicating that they had used a program within the past year.

Immediate and Long-Term Outcomes

Table 6.5 reports the baseline levels of several health and well-being measures for baseline Deployment Life Study families. On average, service members and spouses reported being relatively satisfied with their marriages (means of 4.0 and 3.9 out of five, respectively). Service members and spouses also reported relatively high positive affect (mean scores of 4.2 on the PANAS positive affect scale for both service members and spouses; range is 1 to 5, with higher scores indicating higher positive affect) and relatively low negative affect (scores of 2.0 and 2.1, respectively, on the PANAS negative affect scale; range is 1 to 5, with higher scores indicating higher negative affect) after their conversations and interactions with each other. There was also very little partner marital violence reported by service members and spouses. However, there was discord among some spouses, with about 16 percent of service members and 17 percent of spouses reporting having thought about divorce or separation. Fewer reported actually discussing divorce with their spouses or with a lawyer.

Also, at baseline, few service members and spouses reported problems with psychological or physical health, with the exception of service members having experienced probable TBI. Rates of probable depression were less than 10 percent for service members and spouses, while the rate of probable PTSD for service members was 5 percent. About 8 percent of spouses screened positive for PTSD, although, as noted in Chapter Five, the measure used to assess

Table 6.5
Health and Well-Being of the Baseline Deployment Life Study Sample

Characteristic	Service Member	Spouse
Relationship quality		
Marital satisfaction[a]	4.0 (0.02)	3.9 (0.02)
Positive affect scale, PANAS[b]	4.2 (0.02)	4.2 (0.02)
Negative affect scale, PANAS[b]	2.0 (0.02)	2.1 (0.02)
Partner marital violence[c]	0.9 (0.03)	0.7 (0.03)
Marital dissolution behavior (%)		
Thought about divorce	15.8	17.0
Discussed divorce with spouse	6.0	7.4
Contacting a lawyer about divorce	1.3	0.7
Psychological health		
Probable depression (%)[d]	5.8	7.9
Probable PTSD (%)[e]	4.7	NA
Positive screen for PTSD (%)[e]	NA	7.4
TBI (%)	56.3	NA
Anxiety symptoms[f]	2.1 (0.03)	2.1 (0.03)
Physical health		
Self-rated health[g]	3.8 (0.02)	3.5 (0.03)
Health limitations (%)[h]		
Daily activities	4.6	5.7
Work	5.9	8.5
Social activities	3.7	5.4
Life satisfaction[i]	4.4 (0.02)	4.3 (0.03)
Alcohol and substance use		
Alcohol or drug abuse (modified CAGE)[j]	0.2 (0.02)	0.1 (0.01)
Binge-drinking in past 30 days (%)[k]	27.3	13.8
Off-label prescription drug use, ever (%)	6.3	5.7
Any tobacco use, past 30 days (%)	24.2	19.8
Commitment-to-military scale[l]	4.3 (0.02)	4.5 (0.02)
Satisfaction with military way of life[m]	3.8 (0.03)	NA
Military retention		
Active component		
Retention likelihood or favorability[n]	4.0 (0.04)	4.0 (0.04)

Table 6.5—Continued

Characteristic	Service Member	Spouse
Reserve component (%)		
Remain until retirement	29.5	29.2
Remain until mandatory retirement	54.9	45.0

NOTE: NA = not asked. Values are weighted. Standard deviations are shown in parentheses where applicable. n = 2,724.

[a] Mean of nine items. Range is from 1 to 5, with higher scores indicating greater satisfaction.

[b] Each scale is the mean of four items. Range is from 1 to 5, with higher scores indicating greater positive or negative affect.

[c] Conflict tactic scale is a sum of six items. Range is from 0 to 12, with higher scores indicating more partner violence.

[d] From the PHQ-8.

[e] From the Primary Care Checklist–17 Item (PCL-17) for service members and the PC-PTSD for spouses.

[f] Mean of the four-item anxiety subscale of the Mental Health Inventory–17 Item (MHI-17) (Ware and Sherbourne, 1992). Range is 1 to 6, with higher scores indicating more anxiety.

[g] Responses range from "poor" (1) to "excellent" (5).

[h] The items ask the respondent whether his or her physical health limits his or her ability to function across three domains. For daily activities, the table reports those who say "a lot." For work and social activities, the table reports those who say "moderately" or "quite a bit."

[i] Responses range from "very dissatisfied" (1) to "very satisfied" (5).

[j] Items asked about engagement in alcohol and other substance-use behaviors. Range is from 0 to 7, with higher scores indicating more severe drinking/drug use problems.

[k] The percentage of respondents who binge-drank (5+ drinks for men, or 4+ drinks for women) at least once in the past 30 days.

[l] Mean of three items. Range is 1 to 5, with higher scores indicating higher commitment.

[m] Responses range from "very dissatisfied" (1) to "very satisfied" (5).

[n] Responses range from "very unlikely" (1) to "very likely" (5). For spouses, responses range from "strongly favor service member leaving" (1) to "strongly favor service member staying in" (5).

PTSD among spouses is not directly comparable to the measure used for service members. Mean scores for anxiety are also low (roughly 2 on a scale from 1 to 6). When asked to report possible TBI incidents during their lifetimes, almost 60 percent of service members reported a probable TBI incident. This number is higher than previous reports of service member TBI incident rates (e.g., Schwab et al., 2007; Tanielian and Jaycox, 2008) because the current study asked about possible TBI events experienced over the entire lifetime of the service member, whereas previous research has asked about TBI incidents experienced only during the service member's most recent deployment. Thus, the BTBIS measure used in the current study encompasses TBI incidents that occurred during one or more deployments, as well as incidents that occurred outside of deployment (e.g., car accidents, sports injuries). It should also be noted that the incident rate for TBI requires only that the respondent report having experienced an injury that resulted in an alteration to consciousness, not that he or she currently be experiencing any postconcussive or postinjury morbidity.

On average, service members and spouses rated the status of their physical health as midway between "good" and "very good." About 5 percent of service members and 6 percent of spouses reported that their daily activities were limited "a lot" by their physical health, and between 4 and 9 percent reported that their health interfered with work or social activities.

More than one-quarter of service members and 14 percent of spouses reported binge-drinking in the past 30 days, and 24 percent of service members and 20 percent of spouses reported using tobacco in the past 30 days. Alcohol and drug abuse, as measured by our modified version of the CAGE, is low in this sample (0.2 on a scale from 0 to 7 among service members and 0.1 among spouses). Roughly 6 percent of service members reported ever taking prescription drugs "for the experience or feeling it caused."

Finally, on average, service members and spouses reported that they were "satisfied" with their lives, and service members indicated that they were "somewhat satisfied" with the military way of life. Service members and spouses were moderately committed to the military (e.g., willing to make sacrifices for the military, glad they are part of the military). Similarly, on average, active-duty service members rated their interest in staying on active duty in the military as "likely," and spouses of active-duty service members said that they "somewhat favor" the service member staying on active duty. Reserve component service members were asked to describe their career plans for the reserve. About 30 percent reported that they would stay until they are eligible for retirement, and an additional 55 percent said that they would stay until mandatory retirement. Spouses of reserve component service members reported similar retention intentions, with 29 percent reporting that the service member's retention intentions were to stay in the reserve until retirement, and 45 percent reported that those intentions were to stay until mandatory retirement.

Child Outcomes

As noted earlier, there are two sources of data for children: reports made by the spouse regarding the focal child (n = 1,632) and reports that study children made about themselves (n = 486). According to both sources of data, children in the sample were performing well in school, and few exhibited behavioral or emotional problems. About 65 percent of spouses rated the focal child as getting mostly As and Bs at school, while about 85 percent of study children rated themselves as getting mostly As and Bs in school (see Table 6.6). On average, spouses rated the focal child's schoolwork as "above average" (mean of 4.0 out of 5), as did study children, who rated their own schoolwork as "above average" (mean of 3.8 out of 5). In addition, almost 60 percent of spouses expected the focal child to earn a bachelor's degree, while 48 percent of study children expected to obtain a bachelor's degree. Just over one-quarter of study children expected to pursue a military career when they grow up.

Twenty percent of focal children are scored as having moderate to high emotional difficulties, compared with 28 percent of study children. Spouses rated focal children as being highly prosocial, while study children rated themselves as moderately prosocial. Both focal and study children scored low on anxiety symptoms. Only 65 percent of spouses reported that the focal child had received professional counseling in the past four months if the spouse indicated that the child needed it, suggesting that there may be some unmet need in psychological counseling for children. However, 78 percent of study children reported receiving professional counseling if they themselves indicated that they needed it. Finally, on average, study chil-

Table 6.6
Child Outcomes of the Baseline Deployment Life Study Sample

Characteristic	Focal Child,[a] n = 1,632	Study Child,[a] n = 486
Academic achievement		
Grades: mostly As or Bs (%)	64.8	84.5
Rating of schoolwork[b]	4.0 (0.04)	3.8 (0.05)
Ever repeated a grade (%)	7.1	NA
Educational aspirations (%)		
Less than a bachelor's degree	12.7	12.9
Bachelor's degree	59.8	47.9
Advanced degree (e.g., M.A., Ph.D., M.D.)	27.5	39.2
Expectations for military career (% saying that they expected to have a military career)	NA	26.7
Psychological health		
Moderate to high emotional difficulties (%)[c]	20.4	27.6
Prosocial characteristics[d]	8.5 (0.15)	8.5 (0.07)
Anxiety symptoms[e]	1.5 (0.07)	1.4 (0.1)
If needed, received professional counseling in the past 4 months (%)	65.1	78.2
FES[f]	NA	2.7 (0.02)

NOTE: NA = not asked. Values are weighted. Standard deviations are shown in parentheses where applicable.

[a] Responses for focal children were reported by the spouse. Study children reported their own responses.

[b] Responses range from "failing" (1) to "excellent" (5).

[c] Percentage of respondents scoring moderate (12–15) or high (16–40) on the total emotional difficulties, 4 subscales in the SDQ.

[d] Sum score of the prosocial subscale of the SDQ. Range 0 to 10, with higher scores indicating more prosocial characteristics.

[e] Mean ratings on anxiety symptom scale (SCARED). Range 0 to 10, with higher scores indicating more anxiety.

[f] Mean of six items. Range of 1 to 3, with higher scores indicating better family functioning.

dren reported high scores on the FES, suggesting that they believed their families to be high functioning.

CHAPTER SEVEN

Summary and Conclusion

The Deployment Life Study is the result of a keen interest, on the part of both policymakers and researchers, in military families. Specifically, recent attention by DoD has focused on family readiness. Families, like service members, who are prepared for deployment should, in general, be able to weather the storm. Yet exactly what it means to be "ready" at the family level—what ready families look like, what resources they use, what can help identify vulnerable families—is not well understood. The Deployment Life Study is designed to fill this gap. As such, the overarching goal of this Deployment Life Study is to identify the antecedents, correlates, and consequences of family readiness by collecting longitudinal data from military families across the deployment cycle (i.e., pre-, during, and postdeployment). The study will help to identify which families successfully handle the stress, disruption, and opportunities associated with deployment over time and, using data collected from those families, describe the desirable predeployment characteristics that predict better outcomes among military families. By identifying which families struggle the most with the deployment experience, our work should be able to help DoD target vulnerable families before deployment and design interventions that can build the strength, resilience, and coping mechanisms of these families to help them through the experience. Plus, this information is critical for determining which programs are meeting the needs of families who are most vulnerable, further reducing unnecessary service inefficiencies. It will also allow us to assess the association between deployment and outcomes on whole families, and not simply one family member.

The purpose of this report is to outline the rationale and methods of the Deployment Life Study, describe the baseline sample, and provide a context for future analyses using these data.

The Baseline Deployment Life Study Sample

The Deployment Life Study will follow 2,724 military families (Army, Air Force, Navy, and Marine Corps), both active and reserve, over a three-year period. Interviews, either by Internet or by phone, occur every four months. Up to three household members are interviewed at each wave of data collection: the service member, the spouse of the service member, and a study child (if available). This survey design will collect an unprecedented amount of information about what military families look like and how they handle the obstacles that befall them.

Some initial conclusions about the baseline sample of military families in the Deployment Life Study can be asserted. For example, the majority of our sample service members are male and married to civilian wives. Most are stably married, in first marriages, and have young children. Service members in this sample are experienced, with an average of ten years of service. Thus, most (more than three-quarters) have experienced at least one prior deployment. In absolute terms, low base rates of many known vulnerability factors, in both civilian and military families, exist among the baseline sample—marital conflict and violence, poor family environment conditions, risky behaviors, drug and alcohol use (although binge-drinking may be one exception), and mental health problems. Nonetheless, variability does exist across all of these dimensions, and it, along with changes that will inevitably occur among these families over time and over the deployment cycle, will provide the basis for future analyses. However, it should be reiterated that the baseline alone cannot tell us which families will (or will not) be resilient in the face of a deployment. Only with future waves of data will it be possible to compare within-family differences across time and draw conclusions about what family readiness looks like.

Strengths of the Deployment Life Study

The Deployment Life Study has many strengths that make it uniquely well suited to illuminate family readiness among military families. First, the study design is prospective. Unlike most existing studies of military families, which rely on families to report on their experiences with deployment retrospectively, the Deployment Life Study will assess families not only *while* they experience a deployment but also *before* a deployment. This assessment prior to deployment is critical for disentangling the possible effect of a deployment from the impact of conditions or characteristics that may have existed before the deployment. That is, a prospective design allows us to assess the association between deployment and outcomes among military families, controlling for preexisting conditions.

Second, the Deployment Life Study will be able to assess not only what factors are associated with negative outcomes associated with deployment but also what factors predict successful adaptation to a deployment. This shifts the emphasis from negative to positive, from family vulnerability to family readiness. It also shifts the focus from direct effects of deployment to those factors that alter the effects of deployment: What resources ameliorate (or exacerbate) the impact that deployment can have on military families?

Third, the Deployment Life Study will collect data from not only service members but also their spouses and, in some cases, children. Multi-informant studies can provide a well-rounded understanding of how military families cope with a deployment.

Fourth, whereas many existing studies focus on a single domain (e.g., marital relationships, health and well-being, military experiences), the Deployment Life Study focuses on multiple domains. This allows us to examine processes. That is, we can see whether some domains are affected before others, whether some domains interact with others to produce long-term outcomes, and whether some domains are more sensitive to deployments than others.

Limitations

Despite the strengths of the Deployment Life Study, it is not without limitations that must be kept in mind in any interpretations of the data and results. First, as noted in Chapter Four, a large portion of the families in the original sampling frame were not contacted because of lack of a valid phone number or email. Given frequent moves, contact information is not always kept up-to-date in aggregated personnel records. Military families are a difficult sample to reach. If families that were not contacted are systematically different from those families that were contacted and that are participating in the study, then our sampling procedures may have introduced bias in the baseline sample. Fortunately, there are statistical techniques to correct for this bias. Further, because the study used many measures that are similar to those used in other studies of military families, we will be able to compare and contrast results from the Deployment Life Study with results from those other studies.

Second, as also noted above, the families participating in the Deployment Life Study proved to have, on average, low levels of known risk factors for poor mental health. Moreover, participating families were generally in well-established marriages that had endured multiple deployments prior to the baseline assessment. This suggests that the baseline sample may over-represent those military families that have successfully adapted to military life and (possibly) to prior deployments. Recognizing this possibility, we should point out that, with weighting, the baseline sample will still be representative of the current married, deployable population in the military, which is older and more experienced for the same reasons that the baseline sample is (i.e., younger couples are less likely to remain married and in the military). To the extent that our sample underrepresents younger couples, results using the data may not generalize well to that particular segment of the population. The underrepresentation of the most-vulnerable segments of the population is a common problem in all research that relies on volunteer participation, and it is no less a problem here. Despite these acknowledged limitations, our sample is large enough to allow for analyses that focus exclusively on younger couples, or couples that have not yet experienced a deployment.

Third, the survey asks about sensitive topics, including health and risk behaviors and mental health status, which could introduce bias. As with any survey that includes such topics, the truthfulness of some responses may be in question. Certainly, the research team is taking every precaution to ensure that data are secure and that respondent information is known only to key members of the team. Nonetheless, fear of disclosure may prevent some respondents from being completely truthful.

Fourth, and perhaps the most noteworthy limitation of the Deployment Life Study, is the assumptions regarding causality. Although controlling for family characteristics predeployment may allow for conclusions to be made about the effect that deployment can have on military families, at the same time, if the survey does not contain an exhaustive list of alternative causes (e.g., societal economic trends or personality traits), then there may be a threat to the validity of any causal inferences made from the data. In other words, if we are unable to capture all of the possible reasons for an outcome, then we cannot definitively conclude that deployment is the only reason for the results we find. There are, of course, analytic techniques that can help rule out some of those other factors (e.g., difference-in-differences regression, propensity-score analysis, fixed-effects models), and, as much as possible, they will be utilized in future analyses.

A final word of caution about the Deployment Life Study deals with sample attrition. *Sample attrition* refers to the loss of sample that occurs in longitudinal surveys. However, there are standard techniques for addressing attrition in longitudinal samples (e.g., routine contact during nonsurvey periods, collecting updated contact information) that can be used. It is very likely that some of the families in the baseline sample will not be retained for all of the nine assessments across three years. Nonetheless, the survey design, with repeated measures over time, should provide some leverage on attrition.

Next Steps

Over the next three years, the baseline sample of military families in the Deployment Life Study will continue to be followed, even if the service member separates from the military or the couple divorces. Every effort will be made to maintain contact with families, but, as noted above, some sample attrition in a longitudinal sample of military families is inevitable. In the future, RAND researchers will begin to examine the baseline data, as well as follow-up waves, with more scrutiny, focusing on the three primary research questions that motivate the study: How is deployment associated with the health and well-being of military family outcomes? What accounts for variability across those outcomes? And what behaviors and programs can best buffer military families from the negative impact of deployment? The answers to these, and related questions, will help us to understand what family readiness is and how policymakers and service providers can make sure that every family has an arsenal of resources at their disposal.

References

Abidin, Richard R., *Parenting Stress Index*, 3rd ed., Odessa, Fla.: Psychological Assessment Resources, 1995.

Allen, Natalie J., and John P. Meyer, "The Measurement and Antecedents of Affective, Continuance and Normative Commitment to the Organization," *Journal of Occupational and Organizational Psychology*, Vol. 63, No. 1, March 1990, pp. 1–18.

American Psychiatric Association, *Diagnostic and Statistical Manual of Mental Disorders*, 4th ed., Washington, D.C., 1994.

Anda, Robert F., Janet B. Croft, Vincent J. Felitti, Dale Nordenberg, Wayne H. Giles, David F. Williamson, and Gary A. Giovino, "Adverse Childhood Experiences and Smoking During Adolescence and Adulthood," *Journal of the American Medical Association*, Vol. 282, No. 17, November 3, 1999, pp. 1652–1658.

Angel, Ronald, Linda Burton, P. Lindsay Chase-Lansdale, Andrew Cherlin, and Robert Moffitt, *Welfare, Children, and Families: A Three-City Study*, Ann Arbor, Mich.: Inter-University Consortium for Political and Social Research, ICPSR04701-v7, February 10, 2009.

Ano, Gene G., and Erin B. Vasconcelles, "Religious Coping and Psychological Adjustment to Stress: A Meta-Analysis," *Journal of Clinical Psychology*, Vol. 61, No. 4, 2005, pp. 461–480.

Anthony, Laura Gutermuth, Bruno J. Anthony, Denise N. Glanville, Daniel Q. Naiman, Christine Waanders, and Stephanie Shaffer, "The Relationships Between Parenting Stress, Parenting Behaviour and Preschoolers' Social Competence and Behaviour Problems in the Classroom," *Infant and Child Development*, Vol. 14, 2005, pp. 133–154.

Aranda, Mary Catherine, Laura S. Middleton, Eric Flake, and Beth Ellen Davis, "Psychosocial Screening in Children with Wartime-Deployed Parents," *Military Medicine*, Vol. 176, No. 4, April 2011, pp. 402–407.

Army STARRS—*See* Army Study to Assess Risk and Resilience in Servicemembers.

Army Study to Assess Risk and Resilience in Servicemembers, "Army STARRS Preliminary Data Reveal Some Potential Predictive Factors for Suicide," c. 2012.

Belasco, Amy, *The Cost of Iraq, Afghanistan, and Other Global War on Terror Operations Since 9/11*, Washington, D.C.: Congressional Research Service, RL33110, 2007. As of February 13, 2013: http://www.fas.org/sgp/crs/natsec/RL33110.pdf

Birmaher, B., D. A. Brent, L. Chiappetta, J. Bridge, S. Monga, and M. Baugher, "Psychometric Properties of the Screen for Child Anxiety Related Emotional Disorders (SCARED): A Replication Study," *Journal of the American Academy of Child and Adolescent Psychiatry*, Vol. 38, No. 10, October 1999, pp. 1230–1236.

Birmaher, B., S. Khetarpal, D. Brent, M. Cully, L. Balach, J. Kaufman, and S. M. Neer, "The Screen for Child Anxiety Related Emotional Disorders (SCARED): Scale Construction and Psychometric Characteristics," *Journal of the American Academy of Child and Adolescent Psychiatry*, Vol. 36, No. 4, April 1997, pp. 545–553.

Blais, Ann-Renée, Megan M. Thompson, and Donald R. McCreary, "The Development and Validation of the Army Post-Deployment Reintegration Scale," *Military Psychology*, Vol. 21, No. 3, July 2009, pp. 365–386.

Bliese, Paul D., Kathleen M. Wright, Amy B. Adler, Jeffrey L. Thomas, and Charles W. Hoge, "Timing of Postcombat Mental Health Assessments," *Psychological Services*, Vol. 4, No. 3, August 2007, pp. 141–148.

Blue Star Families, *2012 Military Family Lifestyle Survey: Comprehensive Report—Sharing the Pride of Service*, Washington, D.C., April 27, 2012. As of February 13, 2013:
http://www.bluestarfam.org/Policy/Surveys/Survey_2012

Boscarino, Joseph A., "External-Cause Mortality After Psychologic Trauma: The Effects of Stress Exposure and Predisposition," *Comprehensive Psychiatry*, Vol. 47, No. 6, November–December 2006, pp. 503–514.

Bowen, Gary L., and Dennis K. Orthner, "Postscript: Toward Further Research," in Gary L. Bowen and Dennis K. Orthner, eds., *The Organization Family: Work and Family Linkages in the U.S. Military*, New York: Praeger, 1989, pp. 179–188.

Bradbury, Thomas N., and Benjamin R. Karney, *Intimate Relationships*, New York: W. W. Norton, 2010.

Bray, Robert M., Michael R. Pemberton, Marian E. Lane, Laurel E. Hourani, Mark J. Mattiko, and Lorraine A. Babeu, "Substance Use and Mental Health Trends Among U.S. Military Active Duty Personnel: Key Findings from the 2008 DoD Health Behavior Survey," *Military Medicine*, Vol. 175, No. 6, June 2010, pp. 390–399. As of February 13, 2013:
http://www.dtic.mil/cgi-bin/GetTRDoc?AD=ADA523045

Brener, Nancy D., Janet L. Collins, Laura Kann, Charles W. Warren, and Barbara I. Williams, "Reliability of the Youth Risk Behavior Survey Questionnaire," *American Journal of Epidemiology*, Vol. 141, No. 6, 1995, pp. 575–580.

Brener, Nancy D., Laura Kann, Tim McManus, Steven A. Kinchen, Elizabeth C. Sundberg, and James G. Ross, "Reliability of the 1999 Youth Risk Behavior Survey Questionnaire," *Journal of Adolescent Health*, Vol. 31, No. 4, October 2002, pp. 336–342. As of February 13, 2013:
http://www.cdc.gov/healthyyouth/yrbs/pdf/reliability.pdf

Brewin, C. R., "Systematic Review of Screening Instruments for Adults at Risk of PTSD," *Journal of Traumatic Stress*, Vol. 18, No. 1, February 2005, pp. 53–62.

Brugha, Traolach, Paul Bebbington, Christopher Tennant, and Jane Hurry, "The List of Threatening Experiences: A Subset of 12 Life Event Categories with Considerable Long-Term Contextual Threat," *Psychological Medicine*, Vol. 15, No. 1, February 1985, pp. 189–194.

Bruner, Edward F., *Military Forces: What Is the Appropriate Size for the United States?* Washington, D.C.: Congressional Research Service, 06-RS-21754a, updated January 24, 2006.

BSF—*See* Blue Star Families.

Burnam, M. Audrey, Lisa S. Meredith, Cathy D. Sherbourne, R. Burciaga Valdez, and Georges Vernez, *Army Families and Soldier Readiness*, Santa Monica, Calif.: RAND Corporation, R-3884-A, 1992. As of February 13, 2013:
http://www.rand.org/pubs/reports/R3884.html

Centers for Disease Control and Prevention, "National Health Interview Survey," last updated January 30, 2013, referenced January 7, 2013. As of February 27, 2013:
http://www.cdc.gov/nchs/nhis.htm

Chandra, Anita, Rachel M. Burns, Terri Tanielian, Lisa H. Jaycox, and Molly M. Scott, *Understanding the Impact of Deployment on Children and Families: Findings from a Pilot Study of Operation Purple Camp Participants*, Santa Monica, Calif.: RAND Corporation, WR-566, 2008. As of February 13, 2013:
http://www.rand.org/pubs/working_papers/WR566.html

Chandra, Anita, Sandraluz Lara-Cinisomo, Lisa H. Jaycox, Terri Tanielian, Rachel M. Burns, Teague Ruder, and Bing Han, "Children on the Homefront: The Experience of Children from Military Families," *Pediatrics*, Vol. 125, No. 1, January 1, 2010, pp. 16–25.

Chandra, Anita, Sandraluz Lara-Cinisomo, Lisa H. Jaycox, Terri Tanielian, Bing Han, Rachel M. Burns, and Teague Ruder, *Views from the Homefront: The Experiences of Youth and Spouses from Military Families*, Santa Monica, Calif.: RAND Corporation, TR-913-NMFA, 2011. As of February 13, 2013:
http://www.rand.org/pubs/technical_reports/TR913.html

Chandra, Anita, Laurie T. Martin, Stacy Ann Hawkins, and Amy Richardson, "The Impact of Parental Deployment on Child Social and Emotional Functioning: Perspectives of School Staff," *Journal of Adolescent Health*, Vol. 46, No. 3, March 2010, pp. 218–223.

Chartrand, Molinda M., Deborah A. Frank, Laura F. White, and Timothy R. Shope, "Effect of Parents' Wartime Deployment on the Behavior of Young Children in Military Families," *Archives of Pediatric and Adolescent Medicine*, Vol. 162, No. 11, November 3, 2008, pp. 1009–1014.

Cohen, Sheldon, and Thomas A. Wills, "Stress, Social Support, and the Buffering Hypothesis," *Psychological Bulletin*, Vol. 98, No. 2, September 1985, pp. 310–357.

Commandant of the Marine Corps, "Unit, Personal and Family Readiness Program (UPFRP)," Marine Corps Order 1754.9A, February 9, 2012. As of June 6, 2013:
http://www.marines.mil/Portals/59/Publications/MCO%201754_9A.pdf

Crnic, Keith A., and Mark T. Greenberg, "Minor Parenting Stresses with Young Children," *Child Development*, Vol. 61, No. 5, October 1990, pp. 1628–1637.

De Burgh, Hugo, ed., *Investigative Journalism: Context and Practice*, New York: Routledge, 2008.

De Burgh, H. Thomas, Claire J. White, Nicola T. Fear, and Amy C. Iversen, "The Impact of Deployment to Iraq or Afghanistan on Partners and Wives of Military Personnel," *International Review of Psychiatry*, Vol. 23, No. 2, April 2011, pp. 192–200.

Diener, E., R. A. Emmons, R. J. Larsen, and S. Griffin, "The Satisfaction with Life Scale," *Journal of Personality Assessment*, Vol. 49, No. 1, February 1985, pp. 71–75.

Douglas, Kathy A., Janet L. Collins, Charles Warren, Laura Kann, Robert Gold, Sonia Clayton, James G. Ross, and Lloyd J. Kolbe, "Results from the 1995 National College Health Risk Behavior Survey," *Journal of American College Health*, Vol. 46, No. 2, 1997, pp. 55–67.

Engel, C. C., K. C. Hyams, and K. Scott, "Managing Future Gulf War Syndromes: International Lessons and New Models of Care," *Philosophical Transactions of the Royal Society of London*, Series B: *Biological Sciences*, Vol. 361, No. 1468, April 29, 2006, pp. 707–720.

Farkas, Arthur J., and Lois E. Tetrick, "A Three-Wave Longitudinal Analysis of the Causal Ordering of Satisfaction and Commitment on Turnover Decisions," *Journal of Applied Psychology*, Vol. 74, No. 6, December 1989, pp. 855–868.

Farrell, Albert D., Aleta L. Meyer, and Kamila S. White, "Evaluation of Responding in Peaceful and Positive Ways (RIPP): A School-Based Prevention Program for Reducing Violence Among Urban Adolescents," *Journal of Clinical Child and Adolescent Psychology*, Vol. 30, No. 4, 2001, pp. 451–463.

Felitti, Vincent J., and Robert F. Anda, "The Relationship of Adverse Childhood Experiences to Adult Medical Disease, Psychiatric Disorders and Sexual Behavior: Implications for Healthcare," in Ruth A. Lanius, Eric Vermetten, and Clare Pain, eds., *The Impact of Early Life Trauma on Health and Disease: The Hidden Epidemic*, Cambridge, UK: Cambridge University Press, 2010, pp. 77–87.

Felitti, Vincent J., Robert F. Anda, Dale Nordenberg, D. F. Williamson, A. M. Spitz, V. Edwards, M. P. Koss, and J. S. Marks, "Relationship of Childhood Abuse and Household Dysfunction to Many of the Leading Causes of Death in Adults: The Adverse Childhood Experiences (ACE) Study," *American Journal of Preventive Medicine*, Vol. 14, No. 4, May 1998, pp. 245–258.

Flake, Eric M., Beth Ellen Davis, Patti L. Johnson, and Laura S. Middleton, "The Psychosocial Effects of Deployment on Military Children," *Journal of Developmental and Behavioral Pediatrics*, Vol. 30, No. 4, August 2009, pp. 271–278.

Fredricks, Jennifer A., Phyllis C. Blumenfeld, and Alison H. Paris, "School Engagement: Potential of the Concept, State of the Evidence," *Review of Educational Research*, Vol. 74, No. 1, Spring 2004, pp. 59–109.

Fricker, Ronald D., James Hosek, and Mark E. Totten, *How Does Deployment Affect Retention of Military Personnel?* Santa Monica, Calif.: RAND Corporation, RB-7557-OSD, 2003. As of February 13, 2013:
http://www.rand.org/pubs/research_briefs/RB7557.html

Gibbs, Deborah A., Sandra L. Martin, Ruby E. Johnson, E. Danielle Rentz, Monique Clinton-Sherrod, and Jennifer Hardison, "Child Maltreatment and Substance Abuse Among U.S. Army Soldiers," *Child Maltreatment*, Vol. 13, No. 3, August 2008, pp. 259–268.

Glenn, Norval D., and Kathryn B. Kramer, "The Marriages and Divorces of the Children of Divorce," *Journal of Marriage and Family*, Vol. 49, No. 4, November 1987, pp. 811–825.

Goodman, Robert, "The Strengths and Difficulties Questionnaire: A Research Note," *Journal of Child Psychology and Psychiatry*, Vol. 38, No. 5, 1997, pp. 581–586.

———, "Psychometric Properties of the Strengths and Difficulties Questionnaire," *Journal of the American Academy of Child and Adolescent Psychiatry*, Vol. 40, No. 11, November 2001, pp. 1337–1345.

Goodman, Robert, T. Ford, H. Simmons, R. Gatward, and H. Meltzer, "Using the Strengths and Difficulties Questionnaire (SDQ) to Screen for Child Psychiatric Disorders in a Community Sample," *British Journal of Psychiatry*, Vol. 177, December 2000, pp. 534–539.

Gorman, Gregory H., Matilda Eide, and Elizabeth Hisle-Gorman, "Wartime Military Deployment and Increased Pediatric Mental and Behavioral Health Complaints," *Pediatrics*, Vol. 126, No. 6, December 2010, pp. 1058–1066.

Gutman, Leslie Morrison, and Jacquelynne S. Eccles, "Financial Strain, Parenting Behaviors, and Adolescents' Achievement: Testing Model Equivalence Between African American and European American Single- and Two-Parent Families," *Child Development*, Vol. 70, No. 6, November–December 1999, pp. 1464–1476.

Hill, Reuben, *Families Under Stress: Adjustment to the Crises of War Separation and Reunion*, New York: Harper, 1949.

Hofferth, Sandra, Pamela E. Davis-Kean, Jean Davis, and Jonathan Finkelstein, *The Child Development Supplement to the Panel Study of Income Dynamics: 1997 User Guide*, Ann Arbor, Mich.: Survey Research Center, Institute for Social Research, University of Michigan, 1997. As of February 13, 2013: https://psidonline.isr.umich.edu/CDS/cdsi_userGD.pdf

Hoge, Charles W., Jennifer L. Auchterlonie, and Charles S. Milliken, "Mental Health Problems, Use of Mental Health Services, and Attrition from Military Service After Returning from Deployment to Iraq or Afghanistan," *Journal of the American Medical Association*, Vol. 295, No. 9, March 1, 2006, pp. 1023–1032.

Hoge, Charles W., Carl A. Castro, Stephen C. Messer, Dennis McGurk, Dave I. Cotting, and Robert L. Koffman, "Combat Duty in Iraq and Afghanistan, Mental Health Problems, and Barriers to Care," *New England Journal of Medicine*, Vol. 351, July 1, 2004, pp. 13–22.

Hoge, Charles W., Dennis McGurk, Jeffrey L. Thomas, Anthony L. Cox, Charles C. Engel, and Carl A. Castro, "Mild Traumatic Brain Injury in U.S. Soldiers Returning from Iraq," *New England Journal of Medicine*, Vol. 358, No. 5, January 31, 2008, pp. 453–463.

Hoge, Charles W., Artin Terhakopian, Carl A. Castro, Stephen C. Messer, and Charles C. Engel, "Association of Posttraumatic Stress Disorder with Somatic Symptoms, Health Care Visits, and Absenteeism Among Iraq War Veterans," *American Journal of Psychiatry*, Vol. 164, No. 1, 2007, pp. 150–153.

Homer, *The Odyssey of Homer*, Richmond Lattimore, trans., New York: Harper and Row, 1967.

Hosek, James, *How Is Deployment to Iraq and Afghanistan Affecting U.S. Service Members and Their Families? An Overview of Early RAND Research on the Topic*, Santa Monica, Calif.: RAND Corporation, OP-316-OSD, 2011. As of February 13, 2013:
http://www.rand.org/pubs/occasional_papers/OP316.html

Hosek, James, Jennifer Kavanagh, and Laura L. Miller, *How Deployments Affect Service Members*, Santa Monica, Calif.: RAND Corporation, MG-432-RC, 2006. As of February 13, 2013:
http://www.rand.org/pubs/monographs/MG432.html

Hosek, James, and Paco Martorell, *How Have Deployments During the War on Terrorism Affected Reenlistment?* Santa Monica, Calif.: RAND Corporation, MG-873-OSD, 2009. As of February 13, 2013:
http://www.rand.org/pubs/monographs/MG873.html

Hosek, James, and Trey Miller, *Effects of Bonuses on Active Component Reenlistment Versus Prior Service Enlistment in the Selected Reserve*, Santa Monica, Calif.: RAND Corporation, MG-1057-OSD, 2011. As of February 13, 2013:
http://www.rand.org/pubs/monographs/MG1057.html

Idler, E. L., and R. J. Angel, "Self-Rated Health and Mortality in the NHANES-I Epidemiologic Follow-Up Study," *American Journal of Public Health*, Vol. 80, No. 4, April 1990, pp. 446–452.

Institute of Medicine, *Returning Home from Iraq and Afghanistan: Preliminary Assessment of Readjustment Needs of Veterans, Service Members, and Their Families*, Washington, D.C.: National Academies Press, 2010.

IOM—*See* Institute of Medicine.

Jacobson, I. G., M. A. Ryan, T. I. Hooper, T. C. Smith, P. J. Amoroso, E. J. Boyko, G. D. Gackstetter, T. S. Wells, and N. S. Bell, "Alcohol Use and Alcohol-Related Problems Before and After Military Combat Deployment," *Journal of the American Medical Association*, Vol. 300, No. 6, August 13, 2008, pp. 663–675.

Jacobson, Neil S., and Danny Moore, "Spouses as Observers of the Events in Their Relationship," *Journal of Consulting and Clinical Psychology*, Vol. 49, No. 2, April 1981, pp. 269–277.

Jensen, Peter S., Ronel L. Lewis, and Stephen N. Xenakis, "The Military Family in Review: Context, Risk, and Prevention," *Journal of American Academy of Child Psychiatry*, Vol. 25, No. 2, March 1986, pp. 225–234.

Johnson, David, "Two-Wave Panel Analysis: Comparing Statistical Methods for Studying the Effects of Transitions," *Journal of Marriage and Family*, Vol. 67, No. 4, November 2005, pp. 1061–1075.

Johnson, Jeffrey G., Emily S. Harris, Robert L. Spitzer, and Janet B. W. Williams, "The Patient Health Questionnaire for Adolescents: Validation of an Instrument for the Assessment of Mental Disorders Among Adolescent Primary Care Patients," *Journal of Adolescent Health*, Vol. 30, No. 3, March 2002, pp. 196–204.

Karney, Benjamin R., and Thomas N. Bradbury, "The Longitudinal Course of Marital Quality and Stability: A Review of Theory, Methods, and Research," *Psychological Bulletin*, Vol. 118, No. 1, July 1995, pp. 3–34.

Karney, Benjamin R., and John S. Crown, *Families Under Stress: An Assessment of Data, Theory, and Research on Marriage and Divorce in the Military*, Santa Monica, Calif.: RAND Corporation, MG-599-OSD, 2007. As of February 13, 2013:
http://www.rand.org/pubs/monographs/MG599.html

Karney, Benjamin R., David S. Loughran, and Michael S. Pollard, "Comparing Marital Status and Divorce Status in Civilian and Military Populations," *Journal of Family Issues*, Vol. 33, No. 12, December 2012, pp. 1572–1594.

Karney, Benjamin R., Lisa B. Story, and Thomas N. Bradbury, "Marriages in Context: Interactions Between Chronic and Acute Stress Among Newlyweds," in Tracey A. Revenson, Karen Kayser, and Guy Bodenmann, eds., *Couples Coping with Stress: Emerging Perspectives on Dyadic Coping*, Washington, D.C.: American Psychological Association, 2005, pp. 13–32.

Kiecolt-Glaser, Janice K., and Tamara L. Newton, "Marriage and Health: His and Hers," *Psychological Bulletin*, Vol. 127, No. 4, July 2001, pp. 472–503.

King, Lynda A., Daniel W. King, Dawne S. Vogt, Jeffrey Knight, and Rita E. Samper, "Deployment Risk and Resilience Inventory: A Collection of Measures for Studying Deployment-Related Experiences of Military Personnel and Veterans," *Military Psychology*, Vol. 18, No. 2, 2006, pp. 89–120.

Kline, Anna, Maria Falca-Dodson, Bradley Sussner, Donald S. Ciccone, Helena Chandler, Lanora Callahan, and Miklos Losonczy, "Effects of Repeated Deployment to Iraq and Afghanistan on the Health of New Jersey Army National Guard Troops: Implications for Military Readiness," *American Journal of Public Health*, Vol. 100, No. 2, February 2010, pp. 276–283.

Kroenke, K., R. L. Spitzer, and J. B. Williams, "The PHQ-9: Validity of a Brief Depression Severity Measure," *Journal of General Internal Medicine*, Vol. 16, No. 9, September 2001, pp. 606–613.

Kroenke, K., T. W. Strine, R. L. Spitzer, J. B. Williams, J. T. Berry, and A. H. Mokdad, "The PHQ-8 as a Measure of Current Depression in the General Population," *Journal of Affective Disorders*, Vol. 114, No. 1–3, April 2009, pp. 163–173.

Lara-Cinisomo, Sandraluz, Anita Chandra, Rachel Burns, Lisa Jaycox, Terri Tanielian, Teague Ruder, and Bing Han, "A Mixed-Method Approach to Understanding the Experiences of Non-Deployed Military Caregivers," *Maternal and Child Health Journal*, Vol. 16, No. 2, February 2012, pp. 374–384.

Laurie, H., and Peter Lynn, "The Use of Respondent Incentives on Longitudinal Surveys," in Peter Lynn, ed., *Methodology of Longitudinal Surveys*, Chichester, UK: John Wiley and Sons, 2009, pp. 205–233.

Lester, P., K. Peterson, J. Reeves, L. Knauss, D. Glover, C. Mogil, N. Duan, W. Saltzman, R. Pynoos, K. Wilt, and W. Beardslee, "The Long War and Parental Combat Deployment: Effects on Military Children and At-Home Spouses," *Journal of the American Academy of Child and Adolescent Psychiatry*, Vol. 49, No. 4, April 2010, pp. 310–320.

Lincoln, Alan J., and Kathie Sweeten, "Considerations for the Effects of Military Deployment on Children and Families," *Social Work in Health Care*, Vol. 50, No. 1, 2011, pp. 73–84.

Loeber, Rolf, and Magda Stouthamer-Loeber, "Development of Juvenile Aggression and Violence: Some Common Misconceptions and Controversies," *American Psychologist*, Vol. 53, No. 2, February 1998, pp. 242–259.

Lorig, Kate, Anita Stewart, Philip Ritter, Virginia González, Diana Laurent, and John Lynch, *Outcome Measures for Health Education and Other Health Care Interventions*, Thousand Oaks, Calif.: Sage Publications, 1996.

Loughran, David S., *Wage Growth in the Civilian Careers of Military Retirees*, Santa Monica, Calif.: RAND Corporation, MR-1363-OSD, 2002. As of February 14, 2013:
http://www.rand.org/pubs/monograph_reports/MR1363.html

Loughran, David S., Jacob Alex Klerman, and Craig Martin, *Activation and the Earnings of Reservists*, Santa Monica, Calif.: RAND Corporation, MG-474-OSD, 2006. As of February 14, 2013:
http://www.rand.org/pubs/monographs/MG474.html

Loughran, David S., Paco Martorell, Trey Miller, and Jacob Alex Klerman, *The Effect of Military Enlistment on Earnings and Education*, Santa Monica, Calif.: RAND Corporation, TR-995-A, 2011. As of February 14, 2013:
http://www.rand.org/pubs/technical_reports/TR995.html

Lyle, David S., "Using Military Deployments and Job Assignments to Estimate the Effect of Parental Absences and Household Relocations on Children's Academic Achievement," *Journal of Labor Economics*, Vol. 24, No. 2, April 2006, pp. 319–350.

Mansfield, Alyssa J., Jay S. Kaufman, Stephen W. Marshall, Bradley N. Gaynes, Joseph P. Morrissey, and Charles C. Engel, "Deployment and the Use of Mental Health Services Among U.S. Army Wives," *New England Journal of Medicine*, Vol. 362, January 14, 2010, pp. 101–109.

Martorell, Paco, Jacob Alex Klerman, and David S. Loughran, *How Do Earnings Change When Reservists Are Activated? A Reconciliation of Estimates Derived from Survey and Administrative Data*, Santa Monica, Calif.: RAND Corporation, TR-565-OSD, 2008. As of February 14, 2013:
http://www.rand.org/pubs/technical_reports/TR565.html

Mayfield, Demmie, Gail McLeod, and Patricia Hall, "The CAGE Questionnaire: Validation of a New Alcoholism Screening Instrument," *American Journal of Psychiatry*, Vol. 131, 1974, pp. 1121–1123.

Moos, Rudolf H., and Bernice S. Moos, *Family Environment Scale Manual*, Menlo Park, Calif.: Consulting Psychologists Press, 1994.

Murphey, David A., Kristen E. Darling-Churchill, and Alison J. Chrisler, *The Well-Being of Young Children in Military Families: A Review and Recommendations for Further Study*, Washington, D.C.: Child Trends, January 2011.

National Center for Education Statistics, "National Household Education Surveys Program (NHES)," undated, referenced January 7, 2013. As of February 14, 2013:
http://nces.ed.gov/nhes/

Norwood, Ann E., C. S. Fullerton, and K. P. Hagen, "Those Left Behind: Military Families," in Robert J. Ursano and Ann E. Norwood, eds., *Emotional Aftermath of the Persian Gulf War: Veterans, Families, Communities, and Nations*, Arlington, Va.: American Psychiatric Press, 1996, pp. 163–196.

O'Keefe, R. A., M. E. Eyre, and David L. Smith, "Military Family Service Centers," in Florence Whiteman Kaslow and Richard I. Ridenour, eds., *The Military Family: Dynamics and Treatment*, New York: Guilford Press, 1984, pp. 254–268.

O'Reilly, Charles A., and Jennifer A. Chatman, "Culture as Social Control: Corporations, Cults, and Commitment," *Research in Organizational Behavior*, Vol. 18, 1996, pp. 157–200.

Panel Study of Income Dynamics, undated home page. As of January 7, 2013
http://psidonline.isr.umich.edu

Park, Crystal, Lawrence H. Cohen, and Lisa Herb, "Intrinsic Religiousness and Religious Coping as Life Stress Moderators for Catholics Versus Protestants," *Journal of Personality and Social Psychology*, Vol. 59, No. 3, September 1990, pp. 562–574.

Pavot, William, and Ed Diener, "Review of the Satisfaction with Life Scale," *Psychological Assessment*, Vol. 5, No. 2, June 1993, pp. 164–172.

Pincus, Simon H., Robert House, Joseph Christenson, and Lawrence E. Adler, "The Emotional Cycle of Deployment: A Military Family Perspective," *U.S. Army Medical Department Journal*, April–June 2001, pp. 15–23.

Polusny, M. A., C. R. Erbes, M. Murdoch, P. A. Arbisi, P. Thuras, and M. B. Rath, "Prospective Risk Factors for New-Onset Post-Traumatic Stress Disorder in National Guard Soldiers Deployed to Iraq," *Psychological Medicine*, Vol. 41, No. 4, April 2011, pp. 687–698.

Prins, A., P. Ouimette, R. Kimerling, R. P. Camerond, D. S. Hugelshofer, J. Shaw-Hegwer, A. Thrailkill, F. D. Gusman, and J. I. Sheikh, "The Primary Care PTSD Screen (PC-PTSD): Development and Operating Characteristics," *International Journal of Psychiatry in Clinical Practice*, Vol. 9, No. 1, January 1, 2004, pp. 9–14.

PSID—*See* Panel Study of Income Dynamics.

Public Law 110-181, National Defense Authorization Act for Fiscal Year 2008, January 28, 2008. As of February 15, 2013:
http://www.gpo.gov/fdsys/pkg/PLAW-110publ181/html/PLAW-110publ181.htm

Ramchand, Rajeev, Terry L. Schell, Benjamin R. Karney, Karen Chan Osilla, Rachel M. Burns, and Leah Barnes Caldarone, "Disparate Prevalence Estimates of PTSD Among Service Members Who Served in Iraq and Afghanistan: Possible Explanations," *Journal of Traumatic Stress*, Vol. 23, No. 1, 2010, pp. 59–68.

Rauer, Amy J., Benjamin R. Karney, Cynthia W. Garvan, and Wei Hou, "Relationship Risks in Context: A Cumulative Risk Approach to Understanding Relationship Satisfaction," *Journal of Marriage and Family*, Vol. 70, No. 5, December 2008, pp. 1122–1135.

Reger, G. M., and G. A. Gahm, "Virtual Reality Exposure Therapy for Active Duty Soldiers," *Journal of Clinical Psychology*, Vol. 64, No. 8, August 2008, pp. 940–946.

Reichman, Nancy, Julien Teitler, Irwin Garfinkel, and Sara McLanahan, "Fragile Families: Sample and Design," *Children and Youth Services Review*, Vol. 23, No. 4–5, April–May 2001, pp. 303–326.

Renshaw, Keith D., "An Integrated Model of Risk and Protective Factors for Post-Deployment PTSD Symptoms in OEF/OIF Era Combat Veterans," *Journal of Affective Disorders*, Vol. 128, No. 3, February 2011, pp. 321–326.

Rosen, Leora N., and D. B. Durand, "Marital Adjustment Following Deployment," in James A. Martin, Leora N. Rosen, and Linette R. Sparacino, eds., *The Military Family: A Practice Guide for Human Service Providers*, Westport, Conn.: Praeger, 2000, pp. 153–165.

Rosenthal, Doreen A., and S. Shirley Feldman, "The Influence of Perceived Family and Personal Factors on Self-Reported School Performance of Chinese and Western High School Students," *Journal of Research on Adolescence*, Vol. 1, No. 2, 1991, pp. 135–154.

Rostker, Bernard D., *I Want You! The Evolution of the All-Volunteer Force*, Santa Monica, Calif.: RAND Corporation, MG-265-RC, 2006. As of February 14, 2013:
http://www.rand.org/pubs/monographs/MG265.html

Rubin, Allen, and Helena Harvie, "A Brief History of Social Work with the Military and Veterans," in Allen Rubin, Eugina L. Weiss, and Jose E. Coll, eds., *Handbook of Military Social Work*, Hoboken, N.J.: John Wiley and Sons, 2012, pp. 3–20.

Schell, Terry L., and Grant N. Marshall, "Survey of Individuals Previously Deployed for OEF/OIF," in Terri Tanielian and Lisa H. Jaycox, eds., *Invisible Wounds of War: Psychological and Cognitive Injuries, Their Consequences, and Services to Assist Recovery*, Santa Monica, Calif.: RAND Corporation, MG-720-CCF, 2008, pp. 87–116. As of February 14, 2013:
http://www.rand.org/pubs/monographs/MG720.html

Schumm, W. R., D. B. Bell, and G. Resnick, "Recent Research on Family Factors and Readiness: Implications for Military Leaders," *Psychological Reports*, Vol. 89, No. 1, August 2001, pp. 153–165.

Schwab, K. A., B. Ivins, G. Cramer, W. Johnson, M. Sluss-Tiller, K. Kiley, W. Lux, and D. Warden, "Screening for Traumatic Brain Injury in Troops Returning from Deployment in Afghanistan and Iraq: Initial Investigation of the Usefulness of a Short Screening Tool for Traumatic Brain Injury," *Journal of Head Trauma Rehabilitation*, Vol. 22, No. 6, November–December 2007, pp. 377–389.

Segal, Mady Wechsler, "The Military and the Family as Greedy Institutions," *Armed Forces and Society*, Vol. 13, No. 1, Fall 1986, pp. 9–38.

Serafino, Nina M., *Peacekeeping: Issues of U.S. Military Involvement*, Washington, D.C.: Congressional Research Service, Issue Brief 94040g, updated February 7, 2003.

Sherbourne, Cathy Donald, Ron D. Hays, Lynn Ordway, M. Robin DiMatteo, and Richard L. Kravitz, "Antecedents of Adherence to Medical Recommendations: Results from the Medical Outcomes Study," *Journal of Behavioral Medicine*, Vol. 15, No. 5, October 1992, pp. 447–468.

Smith, Tyler C., Margaret A. K. Ryan, Deborah L. Wingard, Donald J. Slymen, James F. Sallis, and Donna Kritz-Silverstein, "New Onset and Persistent Symptoms of Post-Traumatic Stress Disorder Self Reported After Deployment and Combat Exposures: Prospective Population Based US Military Cohort Study," *British Medical Journal*, Vol. 336, No. 7640, February 16, 2008, pp. 366–371.

Sollinger, Jerry M., Gail Fisher, and Karen N. Metscher, "The Wars in Afghanistan and Iraq: An Overview," in Terri Tanielian and Lisa H. Jaycox, eds., *Invisible Wounds of War: Psychological and Cognitive Injuries, Their Consequences, and Services to Assist Recovery*, Santa Monica, Calif.: RAND Corporation, MG-720-CCF, 2008, pp. 19–32. As of February 14, 2013:
http://www.rand.org/pubs/monographs/MG720.html

Songer, Thomas J., and Ronald E. LaPorte, "Disabilities Due to Injury in the Military," *American Journal of Preventive Medicine*, Vol. 18, No. 3, Suppl. 1, April 2000, pp. 33–40.

SteelFisher, Gillian K., Alan M. Zaslavsky, and Robert J. Blendon, "Health-Related Impact of Deployment Extensions on Spouses of Active Duty Army Personnel," *Military Medicine*, Vol. 173, No. 3, March 2008, pp. 221–229.

Straus, Murray A., Sherry L. Hamby, Sue Boney-McCoy, and David B. Sugarman, "The Revised Conflict Tactics Scales (CTS2): Development and Preliminary Psychometric Data," *Journal of Family Issues*, Vol. 17, No. 3, May 1996, pp. 283–316.

Tanielian, Terri, and Lisa H. Jaycox, eds., *Invisible Wounds of War: Psychological and Cognitive Injuries, Their Consequences, and Services to Assist Recovery*, Santa Monica, Calif.: RAND Corporation, MG-720-CCF, 2008. As of February 14, 2013:
http://www.rand.org/pubs/monographs/MG720.html

Taylor, Shelley E., "Social Support: A Review," in Howard S. Friedman, ed., *The Oxford Handbook of Health Psychology*, New York: Oxford University Press, 2011, pp. 189–214.

Thomas, J. L., J. E. Wilk, L. A. Riviere, D. McGurk, C. A. Castro, and C. W. Hoge, "Prevalence of Mental Health Problems and Functional Impairment Among Active Component and National Guard Soldiers 3 and 12 Months Following Combat in Iraq," *Archives of General Psychiatry*, Vol. 67, No. 6, June 2010, pp. 614–623.

Thompson, Briony M., and Lauren Cavallaro, "Gender, Work-Based Support and Family Outcomes," *Stress and Health*, Vol. 23, No. 2, April 2007, pp. 73–85.

Thomsen, Cynthia J., Valerie A. Stander, Stephanie K. McWhorter, Mandy M. Rabenhorst, and Joel S. Milner, "Effects of Combat Deployment on Risky and Self-Destructive Behavior Among Active Duty Military Personnel," *Journal of Psychiatric Research*, Vol. 45, No. 10, 2011, pp. 1321–1331.

Trail, Thomas E., and Benjamin R. Karney, "What's (Not) Wrong with Low-Income Marriages," *Journal of Marriage and Family*, Vol. 74, June 2012, pp. 413–427.

Under Secretary of Defense for Personnel and Readiness, *Military Family Readiness*, Washington, D.C., Department of Defense Instruction 1342.22, July 3, 2012. As of August 5, 2013:
http://www.dtic.mil/whs/directives/corres/pdf/134222p.pdf

U.S. Army Family and Morale, Welfare and Recreation Programs, "MWR Research," undated web page. As of February 18, 2013:
http://www.armymwr.org/programs/research/default.aspx

U.S. Department of Health and Human Services, Substance Abuse and Mental Health Services Administration, Office of Applied Studies, *National Survey on Drug Use and Health, 2007*, Ann Arbor, Mich.: Inter-University Consortium for Political and Social Research, released December 1, 2008, updated January 4, 2013. As of February 27, 2013:
http://dx.doi.org/10.3886/ICPSR23782.v3

U.S. Department of Veterans Affairs, *Analysis of VA Health Care Utilization Among Operation Enduring Freedom (OEF), Operation Iraqi Freedom (OIF), and Operation New Dawn (OND) Veterans—Revised: Cumulative from 1st Qtr FY 2002 Through 3rd Qtr FY 2012 (October 1, 2001–June 30, 2012)*, Washington, D.C.: U.S. Department of Veterans Affairs, Veterans Health Administration, Office of Public Health, Post-Deployment Health Group, Epidemiology Program, revised December 2012, referenced January 3, 2013. As of February 13, 2013:
http://www.publichealth.va.gov/docs/epidemiology/healthcare-utilization-report-fy2012-qtr3.pdf

U.S. House of Representatives, Heroes at Home Act of 2009, House Bill 667, 111th Congress, referred to the Committee on Armed Services Subcommittee on Military Personnel on February 6, 2009. As of February 14, 2013:
http://beta.congress.gov/bill/111th-congress/house-bill/667

U.S. Senate, Servicemembers Mental Health Care Commission Act, Senate Bill 1429, 111th Congress, hearings held in the Committee on Veterans' Affairs on October 21, 2009. As of February 14, 2013:
http://beta.congress.gov/bill/111th-congress/senate-bill/1429

Van Vranken, E. W., Linda K. Jellen, Kathryn H. M. Knudson, David H. Marlowe, and Mady W. Segal, *The Impact of Deployment Separation on Army Families*, Washington, D.C.: Walter Reed Army Institute of Research, Division of Neuropsychiatry, 1984. As of January 3, 2013:
http://www.dtic.mil/dtic/tr/fulltext/u2/a146595.pdf

Verdeli, H., C. Baily, E. Vousoura, A. Belser, D. Singla, and G. Manos, "The Case for Treating Depression in Military Spouses," *Journal of Family Psychology*, Vol. 25, No. 4, August 2011, pp. 488–496.

Vincent, A. S., J. Bleiberg, S. Yan, B. Ivins, D. L. Reeves, K. Schwab, K. Gilliland, R. Schlegel, and D. Warden, "Reference Data from the Automated Neuropsychological Assessment Metrics for Use in Traumatic Brain Injury in an Active Duty Military Sample," *Military Medicine*, Vol. 173, No. 9, September 2008, pp. 836–852.

Ware, John E., Jr., Mark Kosinski, and Susan D. Keller, *How to Score the SF-12 Physical and Mental Health Summary Scales*, Boston, Mass.: Health Institute, New England Medical Center, 1995.

Ware, John E., Jr., and Cathy Donald Sherbourne, "The MOS 36-Item Short-Form Health Survey (SF-36): I. Conceptual Framework and Item Selection," *Medical Care*, Vol. 30, No. 6, June 1992, pp. 473–483.

Watson, D., L. A. Clark, and A. Tellegen, "Development and Validation of Brief Measures of Positive and Negative Affect: The PANAS Scales," *Journal of Personality and Social Psychology*, Vol. 54, No. 6, June 1988, pp. 1063–1070.

Weathers, Frank W., Jennifer A. Huska, and Terence M. Keane, *The PTSD Checklist Military Version (PCL-M)*, Boston, Mass.: National Center for PTSD, 1991.

Weathers, Frank W., Brett T. Litz, Debra S. Herman, Jennifer A. Huska, and Terence M. Keane, "The PTSD Checklist (PCL): Reliability, Validity, and Diagnostic Utility," paper presented at the annual meeting of the International Society for Traumatic Stress Studies, San Antonio, Texas, October 1993. As of February 14, 2013:
http://www.pdhealth.mil/library/downloads/pcl_sychometrics.doc

Weiss, Robert L., and Mary C. Cerreto, "The Marital Status Inventory: Development of a Measure of Dissolution Potential," *American Journal of Family Therapy*, Vol. 8, No. 2, 1980, pp. 80–85.

Werber, Laura, Margaret C. Harrell, Danielle M. Varda, Kimberly Curry Hall, Megan K. Beckett, and Stefanie Stern, *Deployment Experiences of Guard and Reserve Families: Implications for Support and Retention*, Santa Monica, Calif.: RAND Corporation, MG-645-OSD, 2008. As of February 14, 2013: http://www.rand.org/pubs/monographs/MG645.html

White, Lynn K., and Alan Booth, "The Quality and Stability of Remarriages: The Role of Stepchildren," *American Sociological Review*, Vol. 50, No. 5, October 1985a, pp. 689–698.

———, "The Transition to Parenthood and Marital Quality," *Journal of Family Issues*, Vol. 6, No. 4, December 1985b, pp. 435–449.

Zatzick, D. F., C. R. Marmar, D. S. Weiss, W. S. Browner, T. J. Metzler, J. M. Golding, A. Stewart, W. E. Schlenger, and K. B. Wells, "Posttraumatic Stress Disorder and Functioning and Quality of Life Outcomes in a Nationally Representative Sample of Male Vietnam Veterans," *American Journal of Psychiatry*, Vol. 154, No. 12, December 1997, pp. 1690–1695.

Zohar, A. H., G. Shen, A. Dycian, D. Pauls, A. Apter, R. King, D. Cohen, and S. Kron, "The Military Life Scale: A Measure of Perceived Stress and Support in the Israeli Defense Force," *Israel Journal of Psychiatry and Related Sciences*, Vol. 41, No. 1, 2004, pp. 33–44.